MARK STEYN...

...on Al Gore's face:

Kriss was invited to do the Vice-President's make-up for the first Gore-Bush game. Before the debate, the local papers were running endless profiles of the hometown gal. After the debate, she entered the witness protection program. Many Granite Staters, vaguely aware that a local business had been accorded the honor of showing Al off at his best, assumed the contract had gone to the embalming department of the Lambert Funeral Home on Elm Street.

...on Salman Rushdie's spine:

Roy Hattersley, the Labour Party's deputy leader, attempted to split the difference by arguing that, while he of course supported freedom of speech, perhaps it would be better not to bring out a paperback edition of Rushdie's novel. He was in favour of artistic freedom, but only in hard covers - and certainly, when it comes to soft spines, Lord Hattersley knows whereof he speaks.

...on Jenna Bush's liver:

The only "problem" Jenna has is getting a drink. She can drive, vote, own a house, join the army, get a "civil union" with a lesbian. She can do everything an adult can except wash down her incendiary enchiladas with a margarita. She can buy a gun, shoot up the liquor store and steal the beer. But she cannot walk in and purchase any.

...on Bill Clinton's executive branch:

The founding fathers, in their wisdom, had not foreseen a formal constitutional role for the First Member. That curious innovation has been left to the 42nd President, whose penis has lately had more official engagements than its nominal Commander-in-Chief.

...on the Duchess of York's toe:

Grasp foot firmly with both hands, raise to mouth and then, as the old song says, "Tip Toe Through The Two Lips".

MARK STEYN

can be read regularly in *The Atlantic Monthly, The Chicago Sun-Times,* Britain's *Daily Telegraph* and *Sunday Telegraph, The Irish Times, The Jerusalem Post, National Review, The New Criterion, The New York Sun, The Spectator, The Washington Times,* many other publications around the world, and at SteynOnline.com.

COLLECT THE SET!

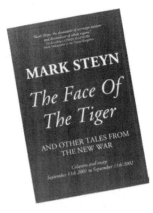

| *The Face Of The Tiger* *And Other Tales From The New War* (2002) | *Broadway Babies Say Goodnight:* *Musicals Then And Now* (1997) |

www.SteynOnline.com

MARK STEYN

From Head To Toe

AN ANATOMICAL ANTHOLOGY

STOCKADE
BOOKS

Published in 2004 by
Stockade Books
CP 843, Succursale H
rue Ste-Catherine ouest
Montréal, Québec
H3G 2M8

Printed and bound in the Province of Québec (Canada)

ISBN 0-9731570-2-X

First Edition

From head...

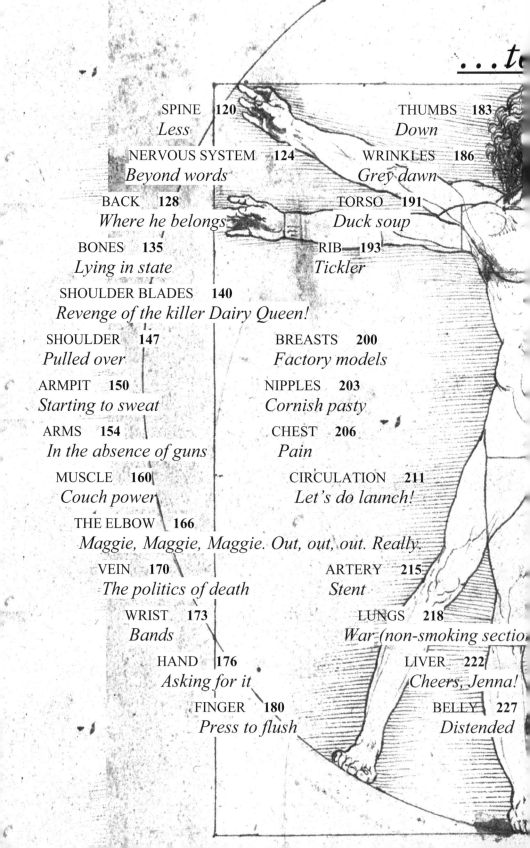

NOTE

These columns originally appeared in the following publications: The American Spectator, The Chicago Sun-Times, *Britain's* Daily Telegraph *and* Sunday Telegraph, *the London* Evening Standard *and* Independent, The Jerusalem Post, *Canada's* National Post, *America's* National Review *and* New Criterion, Slate, The Spectator, The Times *of London, and* The Wall Street Journal. *I would like to thank the editors at the respective titles: R Emmett Tyrrell Jr, Wlady Pleszczynski and Marc Carnegie at* The American Spectator; *Nigel Wade and Steve Huntley at* The Chicago Sun-Times; *Charles Moore, Dominic Lawson, Sarah Sands, Bob Cowan, Sarah Crompton, Christopher Howse and Mark Law at the* Telegraph; *the late John Leese and the late Genevieve Cooper at* The Evening Standard; *Tom Sutcliffe at* The Independent; *Saul Singer and Elliot Jager at* The Jerusalem Post; *Ken Whyte, Martin Newland, Natasha Hassan, John O'Sullivan and Ruth-Ann MacKinnon at* The National Post; *Rich Lowry and Jay Nordlinger at* National Review; *Hilton Kramer and Roger Kimball at* The New Criterion; *Michael Kinsley, Judith Shulevitz and David Greenberg at* Slate; *Frank Johnson, Boris Johnson, Liz Anderson and Stuart Reid at* The Spectator; *Daniel Johnson at* The Times; *Max Boot and Tunku Varadarajan at* The Wall Street Journal.

As a rule, this collection retains the spellings of the originating publication, whether British, American or Canadian. So, if you dislike finding an "s" in the middle of "defence" or an extra "l" in the middle of "marvelous", my marvellous defense is: relax, the word will likely recur with an entirely different spelling two or three pages further on. Or farther on.

HEAD TO TOE

Uneasy lies the crown

June 7th 2001
The National Post

I F, AS THE old saw has it, 90 per cent of journalism is announcing that Lord Jones is dead to people who never knew Lord Jones was alive, much of the remaining ten per cent involves telling people Lord Jones is dead without telling them why. Pity the hapless staff reporter at the government-owned *Rising Nepal* to whom fell the task of writing last Sunday's front-page story. "His Royal Highness Crown Prince Dipendra Bir Bikram Shah Dev has been proclaimed His Majesty the King of the Kingdom of Nepal," he began, uncontentiously enough, and continued:

> *His Majesty the King was born on 13 Asad, 2028 B. S. (June 27, 1971 A.D.) at the Narayanhity Royal Palace as the eldest son of His late Majesty King Birendra Bir Bikram Shah Dev and Her late Majesty Queen Aishwarya Rajya Laxmi Devi Shah... His Majesty the King received his primary school education from Kanti Ishwori Shishu Vidhyalaya, Kathmandu. Having passed the Class 3 district level examinations in the first division, His Majesty the King joined Budhanilkantha High School and passed the School Leaving Certificate Examinations in the First division. Then, His Majesty the King completed 'O' and 'A' levels from Eton College of the United Kingdom...*

On and on the paragraphs roll:

> *...BA at Tri-Chandra College ...Tribhuvan University ...studied Geography at the Masters' Level ...pilot's licence from*

the then Department of Civil Aviation... In 2047 B. S. His Majesty was announced as Colonel-In-Chief of the Royal Nepalese Army...

And so on and so forth, B.S. by B.S:

...Patron of the National Sports Council ...Royal Golf Club ...official visits to Thailand, India, China, Britain ...decorated with the Shubharajya Bhishek Padak, Gaddhi Aarohan Padak...

Alas, pressure of space apparently prevented our doughty hack from mentioning the latest entry to Crown Prince Dipendra's resume or, indeed, how Their late Majesties had come so suddenly to be late: On Friday night, Dipendra gunned down his parents and the rest of his family.

The Rising Nepal had decided to go the Elsinore route. You'll recall that in *Hamlet* the King is murdered by his brother, who then marries the Queen - and everyone at the Danish Court decides it's best, in the interests of political stability, to brush the unpleasantness under the carpet and pretend it's all perfectly normal. By Sunday, the official line was that the Royal massacre had been an "accident" - Crown Prince Dipendra had been showing his relatives his new automatic weapon and it had gone off and killed everyone. Even Polonius might have balked at spinning that one. Dipendra was less Shakespeare's *Hamlet*, and more Frank Loesser's robust precis of the play, as sung by Betty Hutton in the 1949 movie *Red, Hot And Blue*:

He bumped off his uncle
And he mickey-finned his mother
Drove his gal to suicide
Stabbed her big brother
'Cuz he didn't want nobody else but himself should live
He was what you might call uncooperative.

Crown Prince Dipendra was what you might call unco-operative. Six days ago, he was tending bar at King Birendra's regular

4

Friday get-together in a palace billiards room. According to one eyewitness interviewed by *The Times* of London, at about 9pm Dipendra slipped out. According to another account, he'd been drinking, had "misbehaved" with a guest, and been ordered by an angry King to leave. Both versions agree that Dipendra returned shortly after, wearing army fatigues, cap pulled low over his face, with an Uzi submachine gun and an M-16 assault rifle. Without saying a word, he strode up to his father and opened fire, the King dying with a look "of utter astonishment" on his face.

Regicide, such a quaint anachronism, plays awkwardly in modern garb. At Eton, where Prince Dipendra was admired as a "damn good shot," his English school chums called him Dippy. If Katmandu were in Idaho or Mississippi and Dippy had gone to a real public school rather than the British kind, we'd be buried under reams of instant analysis on what precisely made him so dippy. Fox would be running "When Kings Kill". There'd be calls for background checks at gun shows, demands to know what Internet sites he visited, questions about whether his Ritalin and/or Prozac should have been decreased and/or increased.

But Katmandu isn't in Idaho. So, on Sunday, with Dipendra in a coma, the government decided that they would play it strictly as "the King is dead, long live the King." The logic of their position was that, even if Dipendra were never to wake up and resume his duties with the National Sports Council, the government would still be obliged to put his likeness on coins and stamps. They would have turned the entire nation into one of those dysfunctional white-trash families in a Sam Shepard play, where everyone knows there's a baby buried in the yard but no one mentions it.

After criticism in the British and Indian press, the Nepalese government began to rethink, and, by happy coincidence, the killer King died Monday. Maybe, mindful of his duty to the Crown not the man, an obliging surgeon intervened, if only in hastening the inevitable - as Lord Dawson, George V's physician, did when he administered a fatal dose of morphine to his patient in order that His Majesty's death would make the first edition of *The Times* and not have to be announced in the vulgar evening tabloids.

Seemliness is all at Royal Courts, in London as in Katmandu. The new new King – Nepal's third in a week - is King Gyanendra, uncle of last weekend's King Dipendra, brother of last week's King Birendra. He announced a three-man commission to discover the truth of what happened, but the investigation's future is uncertain. You'd think Gyanendra would want to know: He not only lost two brothers, two sisters, a sister-in-law, brother-in-law, cousin, niece and nephew, but his own wife - the new Queen Komal - lies in hospital, badly injured. On the other hand, Gyanendra seems to be unpopular, and the conspiracy theorists want to know how come, when everyone else in the family died, he was out of town, his wife was only injured and his son, Paras Shah, managed to emerge unscathed. Gyanendra and Paras are the last remaining males in the Royal Family, and, while Gyanendra is regarded by the editorialists at *The Daily Telegraph* and elsewhere as a sound chap, calming influence, just what's needed, etc, Paras is apparently a drunken playboy reviled by his future subjects for killing a Nepalese pop star in some sort of hit-and-run accident for which he was never charged. Like his killer King-for-a-day cousin, he has immunity from prosecution.

So I disagree with Alexander Rose in yesterday's *Post* that the smooth succession exemplifies the continuity and stability of monarchy. What happened in Katmandu underscores not monarchy's strength but its vulnerability. Had Gyanendra not chanced to be in India when Dippy opened fire, the Shah dynasty would now be history. Dippy seems to have, as the Americans say, "gone postal" – ie, in the US, a guy like that can gun down his fellow postal workers or burger-flippers, but he's unlikely ever to be afforded the chance to take out the entire ruling elite. But the haphazardness of heredity means that, in a monarchy, the truly madly dippy can find themselves at the heart of the nation's destiny, granted the opportunity not just to go postal but to go regal. And so one unstable prince - Dippy - has now entwined his country's fate with that of another unstable prince -

Paras. The regicidal maniac has delivered his kingdom's future into the hands of the popicidal maniac.

The Royal bloodbath is a sobering reminder that any monarchy is only as secure as its current occupant. According to legend, the first King of Nepal's Shah dynasty, Prithivi Narayan Shah, was about to march on the Katmandu valley when he encountered a holy man. The King offered some curd to the sage, who swallowed it, regurgitated it and offered it back. Understandably enough, His Majesty hurled the regurgitated curd to the ground, though not terribly well, for it landed on his feet. Unfortunately for him, the sage turned out to be the Hindu God Gorakh Nath in disguise who decreed that, because of the King's refusal to swallow, his dynasty would last only ten generations - one generation for each curd-covered toe.

The late King Birendra was the 11th generation. Toe-wise, Nepal is in extra time.

HAIR

Lather, rinse, repeat

December 11th 2002
The Wall Street Journal

JOHN KERRY'S hairdresser continues to make waves in Washington. The news that the Massachusetts senator, Democratic Presidential candidate, Vietnam veteran, Big Ketchup spouse, Vietnam veteran, amateur guitarist, Vietnam veteran and Vietnam veteran gets a $75 coiffure from Cristophe's has riveted the Beltway and distracted from his message. ("As a Vietnam veteran, I know what it's like to wake up in a jungle full of terrifying bangs." "So it was hard to find a decent salon over there?")

To be honest, it's not entirely obvious where the 75 bucks goes. I mean, I haven't seen the back of his head in awhile, so it's possible he has an attractively angled nape. Otherwise, the most likely explanation is that it's 15 bucks for the stuff on top but he pays $30 per eyebrow for some Ann Miller industrial-strength lacquer that freezes them into that permanently furrowed look. For a politician as perpetually concerned as Senator Kerry, this is money well spent. Come the New Hampshire primary, when the candidates are doing their grip-and-grins high atop Mount Washington, Al Gore will be howling in agony as the 200-mile per hour winds rip the chest hair out of his low-cut olive polo shirt and scatter it like confetti over gay weddings in neighboring Vermont, but Mr Kerry's furrowed brow will be as attractively immobile as ever.

On the other hand, it could be confusing if some cranky guy in plaid asks the Senator if he'd like to go out hunting for moose. "Er, no thanks. I'm perfectly happy with my regular styling gel."

The Kerry candidacy is such an obvious disaster waiting to happen that it seems a shame to have to wait for it to happen. It hardly seems possible that Republicans could get lucky enough to have a Democratic primary contest between Al Gore and John Kerry, slugging it out in debate after debate with laboriously self-deprecating gags, one about his woodenness, the other about his lack of warmth:

"I *was* proud. To take *the* lead. In invent*ing* Viagra. Unfortunately. I. Took. One. Too. Many."

"That's nothing. My coiffeuse said, 'Do you want your hair frosted?' I said, 'No, just the rest of me.' Hur-hur."

The reason Al Gore isn't in the White House today is because of the cultural disconnect between him and southern rural white males. Though officially running as a Tennessee farmer, he was perceived as an elite Massachusetts liberal. Replacing him with a real elite Massachusetts liberal seems unlikely to return Tennessee, Arkansas and West Virginia to the fold. Especially an elite Massachusetts liberal committed to raising your taxes.

Now already I can hear Senator Kerry frothing like a vat of Alberto Balsam on Don King's head: "I don't want to *raise* taxes. I just want to repeal the tax cuts you were expecting to get but haven't yet. It's not the same!" To which I say: Whatever, dude. But personally I'd save the hair-splitting for Cristophe's. By the time you've spent 20 minutes explaining why your tax hike isn't really a tax hike, the only two words anyone's going to remember are "tax" and "hike".

And this is where the hair comes in. A lot of solemn Democratic operatives have deplored the Beltway obsession with Mr Kerry's $75 hair care: it's much nothing about a 'do, they say; just another of the media's Drudge-fueled descents into gossip and trivia. True, and that's good enough for me. But, if I have to come up with a highfalutin gloss to justify the story, I'd say it's this: the haircut catches the fancy because it seems to cut to the essence of the Kerry candidacy, whose problem as a whole is that it's over-styled. Platform-wise, every strand feels as if it's been exquisitely combed and parted to the finest calibration. The senator's opposed to the death penalty. Fair enough. A lot of folks have a visceral revulsion at the principle of state execution. But whoa, hang on, no, that's not it. He's not some milksop Dukakis type. Kerry's opposed to the death penalty because it's too wimpy. "Putting somebody to sleep on a gurney", as he puts it, isn't cruel enough for Kerry's tastes. Keep him in jail watching cable TV decade after decade. "That is tough, my friend," says Kerry, not like dying, which – in case he hasn't mentioned it this soundbite – is something he knows a lot about: "I've seen people die and I know what it's like to almost die."

Real men don't fry. Only gutless pansy types let these killers off easy by sending 'em to Old Sparky. This is Kerry's answer to compassionate conservatism: sadistic liberalism.

As the great thespian Sir Donald Wolfit said on his deathbed, "Dying is easy. Comedy is hard." But the comedy in Kerry's campaign is effortless. In this ingenious policy coiffure, the crime strand alone parts to both left and right – and forwards and backwards, too. It turns out the senator *is* in favor of the death penalty, but only for terrorists. And that would be – following Kerry's own logic – because your average al-Qaeda guy deserves a less tough punishment than someone who shoots his wife?

Er, well, that's not important. What's important is that "I, in a war, was prepared to kill in defense of my nation."

I always enjoy the bit at the end of the haircut where the stylist holds up the hand mirror so you can see the back and sides. The trouble with Kerry's policies, as the mirror of the one hand reflects the mirror of the other hand reflects the mirror of the first hand, is that it's all back and sides and no front and center. Bill Clinton could have got away with this approach, but today it seems tonally at odds with the electorate: Bush is certainly not undefeatable, but what is certain is that he won't be defeated by a politician whose gut instinct is to have no gut instincts. Mr Kerry has never held an original position for longer than it took his party's interest groups to put the squeeze on him. The Democrats suffered last month because they were perceived on the central issues of war and national security as, at best, tentative and, worse, opportunist. The senator seems set to expand this losing formula from the war to every major policy area, until the entire Democratic platform has achieved the perfect snapped-seesaw symmetry of his eyebrows.

Even the *Good Morning, Night And Noon, Vietnam* talk falls into this category. If the hair clippings are Drudge's fault, Kerry has only himself to blame for turning his war record into a running joke. How long can he go on any subject before bringing up 'Nam? "Senator, is it true you dye your hair?

"I was ready to die for my country, which is more than a lot of the other side were."

The politics of war is complicated: the media couldn't care less about Bob Dole, a genuine hero who suffered appalling physical injury in a great cause – big deal. John McCain was an incompetent combatant whose brain got fried by the gooks in a lost cause – and the media boomers loved him. Kerry is trying to channel McCain, but he won't pull it off. For a start, there's something a little goofy about a man so convinced his service in Vietnam was morally wrong that he stood on the steps of the Capitol and threw away his medals (well, okay, some other guy's medals) now claiming it as the central, if not sole, event in his resume. You can understand why. Much of the rest – Swiss finishing schools, Dukakis' Lieutenant-Governor – is even less marketable. But again it's slightly out of tune: the 2002 election was a disaster for candidates trying to coast on biography – the Widow Carnahan, Walter Mondale, Max Cleland. The public proved more mature: the personal isn't that political; serving your country in Vietnam is an honorable thing but politically irrelevant if you've got no coherent position on the current war. Indeed, Kerry's latest explanation for his vote against the last Gulf War – "the country was still very divided... I was not against using force. I was against moving so precipitously that we didn't have the consent" – is almost a parody of the modern Democrat's inability to rise above poll-testing.

So what does that leave? If you're in search of bold policy positions, the Kerry message is forget the war and taxes and let's get down to the real issues of real concern to real voters - like "a high-speed rail", which is one of "the things we need to do to excite the economy of our country". If we'd spent more on light rail infrastructure, it wouldn't matter if a President shut down LAX traffic control so he could get a $200 haircut on the runway, because everybody else would be on the 4.07 to Buffalo via Phoenix, Grand Forks, Oklahoma City and Duluth.

This then is the first semi-declared Democratic candidate's strategy: Huff about how your tax hike isn't a tax hike. Talk about cleaning up America's rivers. Keep mentioning Vietnam. Lather, rinse, repeat. If you were to create an animatronic Democrat to exemplify all the most disastrous qualities of the 2002 election – the equivocating, the fundamental unseriousness, the reliance on biography even when no one's interested – it would look an awful lot like John Kerry. His friends are right: the hair is a non-issue. But this is a non-campaign, so what else is there? Or as William Randolph Hearst would say: Remember the mane!

BRAIN

The moron wins again

May 3rd 2003
The Daily Telegraph

WHAT DID YOU make of Dubya in his *Top Gun* get-up landing on the USS Lincoln out in the Pacific?

Opinion on the right seems to be divided. Howard Feinman on MSNBC said it was the "strong horse" speech: you'll recall the late Osama, in his farewell video appearance in 2001, said that "when people see a strong horse and a weak horse, by nature they will like the strong horse". Some, however, thought the stallionesque aspects were excessive: Andrew Sullivan worries that the President appeared "hubristic".

Speaking of horse-like aspects, a not insignificant number of female correspondents have e-mailed me photos of the flight-suited jock and drawn attention to the Presidential lunchbox. Having spent most of the 42nd Presidency discussing the distinguishing characteristics of the executive branch, I think we need to "move on".

Over on the left, however, it's business as usual: "Oh, sure, he 'flew the plane'. But he didn't do the tricky tailhook landing, did he? He flew the easy bits – the straight part in mid-air. Anyone can do that – look at Mohammed Atta. And we're still not winning the hearts and minds in Iraq - did you see that anti-American demo over there the other day? Okay, it was about a tenth of the size of the one in Berkeley, but that just shows you how bad things are going. And don't give me that hung-like-a-stallion stuff. That's just the way they design those army suits, to ease the sexual insecurities of the impotent white American male - see Norman Mailer, at great length. You want a real strong horse? Check out the guy second from the left in the New York City Ballet..."

Meanwhile, a show about a numbskull Prez who thinks he's invading a country called "Iraqistania" looks like being the biggest thing on the West End stage since *Cats*. *The Madness Of George Dubya* is really an example of the madness George Dubya causes in his opponents. Let us take it as read that he is not as verbally fluid as his predecessor, who was positively brimming with fluids. On the other hand, few Democrats are, either. Florida Senator Bob Graham was on TV last Sunday, repeatedly referring to SARS as "SCARS", but no snooty media types made cracks about it.

Speaking as a third-rate hack, I'd say articulacy is greatly overrated. Watching the President fly in, I envied a guy who can control an S-3B Viking. I've been in enough Piper Cubs to figure I'd have a sporting chance if the bad guy shot the pilot and I had to pull the plane out of a tailspin and save me and Pussy Galore. But if it was an F102 Delta Dagger, like Bush flew in the Sixties, me and Pussy would be in big trouble. If Bush, who got a National Guard deferment for Vietnam, is a draft dodger like Clinton (as the lefties charge), he's a dodger of a different order. As to whether it requires a sharper mind to fly an F102 than to write a West End play of lame gags pandering to anti-Bush snobs, I leave to others.

But it's clear now that the Bush moron jokes are indestructible. When all your fondest hopes fail - the Iraqi people turn out to be less Ba'athist than the French, Baghdad isn't Stalingrad, the USAF didn't leave millions of dead kids – it's only natural to retreat to your one great surefire crowd-pleaser: "Shrub" (ha-ha) is an idiot, a "stupid white man", a Texan, a born-again Christian fundamentalist nutbar who would be speaking in tongues if he could string three syllables of gibberish together, and any day now he's sure to say something really dumb again and we can all stand around howling with laughter at the poor boob way out of his league, as a BBC correspondent recently revealed that the British press corps did, listening to the President in the overflow room at Camp David.

But, if I may make a suggestion to my friends on the left, do yourselves a favour and chuck the moron gags. It's insufficient to your needs. In case you still haven't noticed, Bush always winds up getting at least 90 per cent of everything he wants, and it can't all be dumb luck. A year ago the President told Trevor McDonald, "I made up my mind that Saddam needs to go." Well, Saddam's gone. In between came a lot of entertaining diplomatic dances in national costume, but, like the third act of *The Nutcracker*, they didn't impact on the plot: in the end, the nut got cracked. Some of his allies - the Prime Minister of Britain - have overcome their squeamishness to regime change. Some of his opponents - the Prime Minister of Canada - were still objecting to regime change even after the regime had changed. But it was Bush's position that counted: one of his strengths is that he won't sacrifice the objective to the process. By contrast, it wasn't always apparent that his predecessor had objectives: what exactly was the desired end when Mr Clinton bombed that aspirin factory in the Sudan? In foreign policy, Clinton had tactics, not strategy: his inability to reach what Ken Starr called "completion" extended far beyond Monica's gullet. On his tax cuts, on missile defence, on Saddam, Bush is completion-focused. That's why, two years in, the Taliban are history, Saddam's out of power, and the Democrats took an historically unprecedented pasting in the mid-term elections. There are reasons to criticise this President – we on the right have a zillion of them – but his stupidity is not one of them.

The chaps who dismiss Bush as a moron forget that what counts is not what a guy does when he's not talking. It's true that he didn't know the name of the leader of Wackistan before he became President. But one advantage of that is that he isn't the prisoner of his past the way, say, Chirac, Schroeder and Putin are. Chirac the sleazy deal-maker, Schroeder the Sixties anti-American peacenik and Putin the KGB hardman seem incapable of rising above their CVs. That's the subtext of the Russian President's extraordinary performance with Blair the other day. How would you feel if you were Putin? Your guys kill more people in a single Moscow theatre than Bush's do liberating

Baghdad. Bush wraps Iraq up in a month, while you've spent years killing hundreds of thousands and reducing Grozny to rubble and your boys are still coming home in boxes.

I'd say *The Madness Of Vlad & Jacques* would make a good play, but no doubt there's no audience for it. Between now and January 2009, whatever Bush does he'll always be a dummy to the smart set. So the "strong horse" can fly a jet across the Pacific? Big deal. You can take a horse to water but you can't make him think.

SKULL

Headhunting

July 14th 2001
The Daily Telegraph

Richard Caborn Knows Nothing About Sport, So What's New?
He's Only The Sports Minister. – The Daily Mail

A man who does not believe in God is to become head of the
BBC's religious broadcasting. Divorcee Alan Bookbinder, 45,
who currently specialises in making science programmes, describes
himself as an 'openhearted' agnostic and lives with his partner. –
The Daily Mail

"AH, COME IN, Mark," said Vanessa, my careers consultant, when I arrived for our weekly session. "We're going to try a dummy 'job interview'. I hope you've done your homework."

"Yes, ma'am!" I said, eager to sound eager.

"So, okay, I'm Lord Rothermere and you're applying for the editorship of *The Daily Mail*."

"Got it."

"First off, what makes you think you'd be right for the job?"

"Well," I said, trying not to sound as if I'd been practising all night in front of the mirror, "I'm highly motivated and a self-starter, your lordship."

"I see," said Vanessa, coolly. "And have you any experience?"

"Oh, lots," I said. "I've worked in newspapers for years. I'm currently a columnist with the *Tele*... I mean, I'm currently 'Senior Executive Columnist' at *The Daily Telegraph*." I gave Vanessa a conspiratorial wink. "I'm not really a senior executive columnist, but, like you said, it's the sort of thing nobody checks. Here, what do you think about these fake business cards I had made up? There's a coin machine in the arcade at Victoria Station that'll do you a hundred for a fiver. They've spelt *Telegraph* with two g's, but I can fix that with the Tipp-Ex."

Vanessa hurled them across the room, sending dozens into the ceiling fan, which instantly shredded them and rained them down on my head, the cruel confetti of dashed hopes. "This is pathetic!" she snapped. "The *Mail* would never hire you in a thousand years!"

"But I'm doing everything you told me to," I protested. "Sound motivated, pad the CV..."

"Don't you read the papers?" she said, and tossed me a couple of cuttings. One read: "New Minister Of Sport Knows Nothing About Sport". The other began: "The BBC's new Head of Religious Broadcasting is a divorcee who doesn't believe in God, lives with a 'partner', hasn't had his children baptised and has never shown any interest in religion."

"The rules have changed, baby," said Vanessa. "Employers don't want self-starters any more; they want non-starters." She buzzed her secretary. "You can send Pedro in now." The door opened and a man in spangled tights, wearing dark glasses and holding a white cane, came wobbling in on a unicycle, crashed and fell in Vanessa's lap.

She struggled out from under. "You're looking at the new chief neurosurgeon of the Royal London Hospital: Pedro the Blind Unicyclist. I must have seen 50 candidates - all well qualified, lots of experience, hotshot docs from Dublin, South Africa, you name it. But somehow they all seemed a bit predictable, a bit safe."

"But surely a neurosurgeon ought to be safe," I said.

"Mark, it's all about thinking outside the box. The punters today are hip, ironic, post-modern. They don't want to go into hospital and spend hours with some stuffy by-the-book type hung up on surgical procedure and sterilised instruments. Pedro brings a fresh new outlook to the operating theatre. He's never been in one, he's not interested in hospitals, he's just what they were looking for."

I was still digesting this when Norman Tebbit burst in. "Great news, Norman," she said. "*Gay Times* has offered you the editor's job."

I was stunned. "But you're not gay," I said to Lord Tebbit, "although, come to think of it, you've always seemed a bit camp."

"I'm not a poof, you jumped-up little media fairy!" he yelled, and nutted me in the forehead. My head began to swim, though that might have been because Pedro was unicycling round the coffee table.

"We've prepared a press release," said Vanessa. "*Gay Times* today announced the appointment of Norman Tebbit as Editor-in-Chief, the first non-gay editor in the paper's history. 'We believe Lord Tebbit brings an exciting and distinctive perspective to gay issues,' said the publishers in a statement. Lord Tebbit himself added: 'Although not a practising gay, I have worked with several gay people in my previous job in the Conservative Party, though I always made sure to sit down the other end of the Cabinet table.'"

"This is rubbish!" I scoffed. "People are being appointed to jobs they've no interest in, that they're not qualified for, that they frankly despise. It's ridiculous. You wouldn't put up with it in your profession, would you?"

"What do you mean?" said Vanessa.

"Well, you're a top London executive headhunter. What job did you do before this one?"

"I was a headhunter."

"Exactly," I said. "You followed a logical progression in your own career."

"No, no, you misunderstand me," said Vanessa. "I was a headhunter. In Borneo." She pulled out the top drawer of her filing cabinet and inside were 14 shrunken skulls. "I'd been doing it 20 years," she sighed, "and I was burnt out. It's a jungle out there. So I was ready for a mid-life career shift, and, frankly, careers consulting has never been easier."

She looked at her watch. "I'd love to stay and chat, but our Job Interview Masterclass is about to start. Ken Clarke and Michael Portillo will be giving lessons in patronising ennui and thinly veiled distaste. Oh, and you're due at an interview. Pedro's old circus is looking for a new unicyclist."

FACE

The mask drops

October 15th 2000
The Chicago Sun-Times

THIS ELECTION has always been about one thing, and it was thoughtfully touched on by Kriss Soterion, the former Miss New Hampshire and owner-operator of Kriss Cosmetics & the Studio of Holistic Beauty in Manchester, New Hampshire. Kriss, you may recall, was invited to do the Vice-President's make-up for the first Gore-Bush game in Boston.

Before the debate, the local papers were running endless profiles of the hometown gal on the eve of her big break. After the debate, she entered the witness protection program. The pancake she smeared on Al's cheeks, on top of his overly worked-out, 23-inch neck, on top of his lumpy suit, combined to make him look like Herman Munster doing a bad Ronald Reagan impression. Many of us Granite Staters, vaguely aware that a local business had been accorded the honor of showing Al off at his best, assumed the contract had gone to the embalming department of the Lambert Funeral Home on Elm Street.

Poor old Kriss would like to make up for her make-up, but she fears she'll never get another chance, and now seems to be going through some existential crisis, riddled with self-doubt and questioning her calling. She told *The New Hampshire Sunday News* that her catastrophic touch-up of Al has caused her to "think deeply" about "the psychology of make-up."

"It just makes me think about the whole thing, about wearing masks," she said. "It's kind of a fascinating subject, to analyze why we hide behind it in the first place."

20

This is not a subject the Vice-President wants to discuss at this stage in the election cycle.

Still, Kriss has a point. For the second presidential face-off, Al was appearing without his face on - not just the foundation and rouge, but in a broader sense: The mouth wasn't merely in non-sighing mode, it was zipped up and hung slack; the eyes seemed dead. The only remnants of last week's Gore were the eyebrows (NB, Kriss: nice pencil liner) imperiously arched with the amused contempt of an overthrown king sitting through his own show trial. On Wednesday night, the Vice-President had the look of a man who'd run out of masks. After the expansive array of dazzling new Gores of the last year, the Vice-President apparently opened his closet and found that his housekeeper had sent all his identities to the cleaners.

Dubya, for his part, connected the dots between personal character and public policy in a subtle but devastating way. When he spoke of an America that was "humble, but strong" as opposed to one that was "arrogant" and thought it could be "all things to all people," he wasn't really talking about the nation per se so much as its embodiment in the commander-in-chief. Al's sneering disdain for Dubya in the first debate isn't just personal distaste but emblematic of his arrogance in government: We'll give you tax credits - but only if you live your life the way we say, from cradle to grave, from pre-school child care to seniors' health plans. And just because we're running your lives doesn't mean we won't be running the rest of the world, too.

Al made the astonishing assertion - perhaps as a result of watching NBC's Olympics coverage - that the rest of the world wants to be "more like America." Many people around the planet admire the United States, but are proudly Irish, Australian, Kazakh or Nepalese, and wish to build their countries their way. When Dubya talked about a humble but strong America, he sounded as if he was talking about himself. When Al talked about a uniquely powerful America imposing its values on everyone else, he was talking about himself, too.

After the first debate, the *Sun-Times'* Bob Novak declared that Gore looked "too big." Massachusetts Dem Barney Frank mocked the observation. "I do not think he can shrink between now and the next debate," he said. "His size will probably remain a constant. Crouching maybe would be about it."

But on Wednesday something happened. Gore did shrink, and shrivel - before our very eyes.

EARS

Beyond politics

August 2nd 2003
The Spectator

WHAT HAPPENED to Liberia? Only three years ago, things were going swimmingly, at least according to President Charles Taylor's Ministry of Information:

We say "well done" to Mr President, and advise him to always keep the communication highway free and clear of any hindrance, so that a people-to-leader and leader-to people approach can be adopted and maintained, so that everyone will at least have the opportunity to have the ears of the Chief Executive, instead of a select few.

By contrast, in 1990 only a select few got the opportunity to have the ears of the then Chief Executive. Samuel Doe had fallen into the hands of Prince Johnson, one of Charles Taylor's allies in the battle to unseat him. Johnson had President Doe stripped to his underpants and then barked into the camera: "That man won't talk! Bring me his ear!" The cameraman did a jerky about-face in time to catch Johnson's guys holding down the President and slicing off his left ear.

"Now the other ear," ordered Johnson. "The right ear." So the boys removed the right one. Then they made the President eat them. But the lads kept the best bits for themselves. They removed His Excellency's genitals and then fought over them, in the belief that the "powers" and "manhood" of the person whose parts you're eating are transferred to the eater.

Times change, and it's now President Taylor's lunchbox on the menu. He's currently trying to avoid becoming just another ear-today-

23

gone-tomorrow Liberian head of state. His former ally, Prince Johnson, has since fallen out with Taylor, relocated to Lagos, been ordained by the Christ Deliverance Ministry, and had a tearful reconciliation with Samuel Doe's widow at the Synagogue Church of All Nations. He now regrets the whole ear-slicing thing and the good news is he's ready to come back and serve his country. So is his fellow warlord Roosevelt Johnson (no relation).

Currently, and somewhat improbably, Liberia has the ear of George W Bush. With Iraq, there was no agreement on what the thing was about: it's all about oil, said the anti-war crowd; it's about the threat Saddam represents to the world, said the pro-crowd. But with Liberia there's virtually unanimous agreement: the US has no vital national interest in the country; its tinpot tyrant is no threat to anybody beyond his backyard; the three warring parties are all disgusting and none has the makings of even a halfway civilized government. For many on the right, these are reasons for steering clear of the place. For the left, they're why we need to send the Marines in right now.

Why would the left, currently berating Bush for "putting our brave men and women in harm's way" in Iraq, be so anxious to put them in harm's way in Liberia? It's precisely the lack of any national interest that makes it appealing to the progressive mind. By intervening in Liberia, you're demonstrating your moral purity. That's why all the folks most vehemently opposed to American intervention in Iraq – from Kofi Annan to the Congressional Black Caucus – are suddenly demanding American intervention in Liberia. *The New York Times* is itching to get in:

> *Three weeks have passed since President Bush called on the Liberian president, Charles Taylor, to step aside, and pledged American assistance in restoring security. But there has been no definitive word here on how or when.*
>
> *"Oh God, oh God, what do we do now?" wailed one woman in front of the entrance to the embassy... A man yelled, "Why can't the Americans come in to rescue us?"*

Three weeks! And Bush is still just talking! The *Times* spent 14 months deploring the "rush to war" in Iraq, but mulling Liberia for three weeks is the worst kind of irresponsible dithering.

Likewise, Democratic Presidential frontrunner Howard Dean of Vermont. "I opposed the war in Iraq because it was the wrong war at the wrong time," says Governor Dean. But Liberia's the right war any time: "Military intervention in Liberia represents an appropriate use of American power." And unlike that desert mess, the Liberian jungle won't be one of them quagmires. Dean confidently predicts that US troops would "stabilise the situation and remain in Liberia for no more than several months."

It makes sense to Frank Griswold, Presiding Bishop of the Episcopal Church – ie, America's Anglicans. On Iraq, he advocated endless jaw-jaw – "diplomatic and multilateral initiatives… a foreign policy that seeks to reconcile and heal…" – but for Liberia he's got the whiff of cordite in his nostrils, demanding immediate "peace keeping forces to end the hostilities and achieve a cease-fire. Only then can an orderly transition be made from the current chaos to a legitimate and stable government."

The New York Times managed to find at least one Liberian who's got the message: "One young man held up a torn sheet of cardboard, his fury scrawled with a black marker. 'G. Bush Killer Liberia,' it said." Wow! Forget Doe and Taylor and the various Johnsons: G. Bush Killer Liberia! Before a single warlord's genital has slipped down Dubya's gullet!

I'm an imperialist and right now no-one could use a little imperialism more than Africa. The British insertion into Sierra Leone was a good thing; Ivory Coast is on balance better off with the French on the ground. Why shouldn't the Americans also have a little piece of the West African mosquito swamp? If a couple of thousand Marines can "stabilise" Liberia, for a great power to deny them seems, as William F Buckley put it, "parochial". But the idea that the US would be there for "no more than several months" and hand over to a "legitimate and stable government" is ludicrous. If the Yanks are there for only a few months, the warlords will keep their ears close to the ground and bide their time. The intervention would be an intermission, after which the show would resume, as it has done after previous desultory interventions in the region.

When advocates of dispatching the Marines say Liberia's a small manageable nation of only three million people, they're making the mistake of looking at the map. That Liberia doesn't exist. The three contiguous West African nations in which the west has been called on to intervene have jumped, decisively, the borders drawn for them by 19th century Europeans. Taylor is credited with having displaced not just (at one time or another) the entire population of his own country but also a significant chunk of the surrounding states'. Liberia's only significant export to its neighbours is chaos. As early as a decade ago, 400,000 Liberians had fled to Sierra Leone, and 100,000 Sierra Leonians had fled to Liberia. In the course of the Nineties, over one million Liberians and Sierra Leonians fled to Guinea and Ivory Coast. Next, half a million Ivorians fled to Guinea and other neighbouring countries when things went belly up there.

Some of those Sierra Leonians in Liberia would like to flee back to Sierra Leone, and some of those Liberians in Sierra Leone would like to flee back to Liberia, and some of those Liberians in Ivory Coast fancy a change of displacement to Sierra Leone and some of those Ivorians in Guinea are minded to check out the displaced persons' scene in Burkina Faso. But if Howard Dean thinks this is a little light six months of "peacekeeping", maybe he should volunteer for the Paul Bremer role.

Charles Taylor, a fellow most Americans had never heard of till a month ago, was educated in Boston. Most of the drunken teen thugs underneath him weren't educated anywhere. And so, when interventionists argue that the leaderships of the various factions are exhausted and ready for a break, the question is whether the gun gangs they nominally control are also in the mood for a sabbatical. In the sprawling cities of West Africa, for the swollen population of unemployed and unemployable illiterate male youths, stealing and killing are pretty much the only rational career choices. In Liberia, male life expectancy in the last five years has declined from 56 to 44 years; in Sierra Leone, it's down to 32. Village life has drained away to the coastal shanty megalopolis, where crime and disease fill the civic and cultural vacuum.

The Congressional Black Caucus blames all this on the legacy of colonialism, but it would be more accurate to call it the legacy of post-colonialism or prematurely terminated colonialism. The first generation of the continent's leaders were those LSE-educated Afro-Marxists who did such a great job at destroying their imperial inheritance. By the time that crowd faded from the scene, the Cold War was over and nobody needed African puppets. So today West Africans find themselves in a land beyond politics. You can't seriously talk of these factions as being Marxist or Maoist or Blairite. They're gangs. There's the Ear-Slicers and the New Ear-Slicers and the Ear-Slicing & Genital-Sautéing Coalition, but none represents any coherent political platform. The video of Samuel Doe's sudden loss of hearing pre-dates the equivalent scene in *Reservoir Dogs* by a couple of years, but that's the valid comparison: these are criminal operations, not political ones. The only difference is that the ear-slicing of Sam Doe wasn't accompanied on the soundtrack by "Stuck In The Middle With You". That'll be left for the US Marines to sing.

To most people in Britain, colonial Africa isn't that long ago. It's only a little over three decades since the Queen was Sierra Leone's first post-independence head of state. But, in a land where male life expectancy is 32, who remembers the late Sixties? Who remembers District Commissioners and functioning schools and non-psychopathic police forces? These are cultures that, except for a few quaintly revived traditions like genital eating, exist more and more completely in a present-tense dystopia. In *The Atlantic Monthly* a few years back, casting around for a phrase to describe the "citizens" of such "states", Robert D Kaplan called them "re-primitivised man". Demographic growth, environmental devastation, accelerated urbanization and civic decay have reduced them to a far more primitive state than their parents and grandparents.

There are signs some Africans understand this. In January, *The East African*'s Charles Onyango-Obbo wrote a column musing on the resurgence of cannibalism, after the UN had reported that Ugandan-backed rebels in the Congo were making their victims' relatives eat the body parts of their loved ones. "It also makes the point," he continued, "that while colonialism is bad, the coloniser who arrives by plane, vehicle, or ship is better - because he will have to build an airport, road, or harbour - than the one who, like the Ugandan army, arrived and withdrew from most of eastern Congo on foot."

Just so. If you're going to be a subject, you might as well be a subject of the most advanced society available. The would-be "liberators" of Liberia, backed by Taylor's enemies in neighbouring regimes, more or less guarantee that the country's future will be as poor and vicious and diseased as they are.

So the question for the Americans is not whether you want to send 2,000 boys in to get picked off for a few months until whichever warlord is willing to be bought can be installed as head of a provisional government after a token "election" for the benefit of the international community (Taylor held his in 1997). The question is whether you want to commit yourself to fixing West Africa.

I know how most Americans would answer that. But the Bush Administration thinks more about the Dark Continent than its predecessor did. Disease in Africa, for example, has been identified as a potential national security threat. An American diplomat recently described the war on terror to me as a Saudi civil war that the Saudis had successfully exported to the rest of the world. What would it take to export West Africa's troubles to the world? For some no-account nickel'n'dime operator, Charles Taylor has done a grand job of destabilizing a region: the French are justifiably miffed that Ivory Coast, one of the least ghastly African states, has been sucked into the same death spiral as its Anglophone neighbours. Where's next? Benin? Togo? Nigeria? If you don't think West Africa can be contained, it'll have to be cured, and that's a 30-year project. Otherwise, George F Kennan's argument against intervention in Somalia holds for the west of the continent, too:

> *The fact is that this dreadful situation cannot possibly be put to rights other than by the establishment of a governing power for the entire territory, and a very ruthless, determined one at that. It would not be a democratic one, because the very prerequisites for a democratic political system do not exist among the people in question.*

On the other hand, if anyone in the Bush Administration were to start talking about Liberia in those terms, you can pretty much guarantee that Howard Dean, Bishop Griswold, and all the other enthusiastic interventionists would be marching up and down chanting, "It's all about diamonds!"

Stay tuned. Or, as they say in Monrovia, keep your ears peeled.

TEMPLE

Shirley, you can't be serious

March 18th 2002
The National Post

ON FRIDAY, SHIRLEY Temple Black, the beloved former child star of "On The Good Ship Lollipop" and "Animal Crackers In My Soup" fame and more recently US Ambassador to Czechoslovakia, pulled out of Liza Minnelli's wedding. Her reason for cancelling at 24 hours' notice was that she was "weirded out" by Liza's fiancé's vast collection of Shirley Temple memorabilia.

That's all I can tell you, I'm afraid. What a tantalizing item, one of thousands of bewildering details swept along in a torrent of verbiage from the New York and Fleet Street tabs, who collectively seem to have decided that, if we don't pull out the stops for "Liza's Nutty Nuptials" (*The New York Post*), then the terrorists will have won!

Liza, for her part, was happy to oblige them, inviting gossip columnist Cindy Adams, the duchess of dish, to be one of her bridesmaids, along with Mia Farrow, Petula Clark, Gina Lollobrigida, Chaka Khan, Esther Williams, an actress from the BBC soap opera "EastEnders" and some eight other stars in various degrees of eclipse. Everyone who was anyone was there, no matter how long ago it was they were anyone: Donna Summer, Mickey Rooney, the Doobie Brothers, Joan Collins, Little Anthony and the Imperials, Jill St John… Even the hotter guests had been on ice for a good quarter-century. But, if you were an MGM contract player in the Forties, had a couple of doowop hits in the Fifties, did some Bob Hope USO tours in 'Nam in the Sixties, were a gay disco diva for 18 months in the Seventies, wore shoulder-pads on an Eighties soap, chances are you had a premium pew to see Liza walk down the aisle on Saturday. The fourth-time bride wore white (low-cut, off the shoulder), the bridesmaids wore black.

I first heard about the wedding some time before Christmas, when it was announced that Liza and her groom, Wossname, David Gest, would be married at St Patrick's Cathedral with Michael Jackson as best man, Elizabeth Taylor as matron of honour and Whitney Houston doing "Here Comes The Bride" - or as Whitney would say, "He-e-e-e-ere Co-o-o-o-omes The-e-e Bri-i-i-i-ide." In the end, as is traditional, Whitney didn't show, telling a tearful, distraught Liza that she was tied up in the studio remixing. Sources say that, privately, she was concerned that she would look ridiculously out of place among a collection of oddballs and has-beens. Hmm. A spokesman for the couple released the following statement: "David and Liza hope Whitney is well."

But Jacko was there. He and Tito Jackson served as David Gest's co-best men, with the remainder of the Jackson Five (Marlon, Randy, Jermaine) as groomsmen. Liz Taylor also showed, though the ceremony was postponed because she turned up in her slippers, so someone had to go back to the hotel to get her shoes. It wasn't St Pat's, but another Fifth Avenue church, less fussy about Liza and David's variations on the traditional ceremony: After walking down the aisle, the bride was serenaded at the altar by Natalie Cole singing "Unforgettable". It was obvious to all that, whatever the rumours, the groom preferred his betrothed to the three Liza drag queens who had serenaded him with a medley of her hits at his bachelor party the night before.

Mr Gest, a producer of all-star galas, was said to have been subdued at his farewell to bachelorhood, and left early. He had various last-minute problems: The security team quit in a dispute over remuneration, and several stellar invitees backed out after hearing that he'd sold exclusive picture rights for the occasion to the British gossip magazine *OK!* He had to contend with catty asides from the likes of Elton John: Asked what he'd like to give Liza as a wedding present, Sir Elton replied, "A heterosexual husband." Nonetheless, nothing could dim the lustre of what *The New York Post* described as "the grandest wedding since Charles and Diana." "The interest in this wedding has been incomparable," observed Cindy Adams. "Mrs Anthony Quinn came specially from Rhode Island."

To be honest, I was a bit disappointed not to get an invite. I believe I'm the only living Canadian to have met all three of Liza's ex-husbands: the gay one; the one whose dad was the Tin Man in *The Wizard Of Oz*; and the one who was big in the art world and so unshowbizzy that he stayed in the kitchen and slipped out the service door the last time I was at Liza's pad. But a mutual friend says this wedding was the groom's production. She'd met him a few days before September 11th at Michael Jackson's 30th anniversary celebrations, which he produced. A Jacko spectacular is probably not the best place to meet a husband, but, having said yes, Liza seemed to feel she owed it both to him and to her public to turn her wedding into a performance. As the *Sunday Mirror* headline put it, "Wife Is A Cabaret."

Alas, the heavyweight titles fretted about the significance of the event: "Didn't someone say that America had moved on from its celebrity-culture days after last September?" scoffed Britain's *Independent*. It's increasingly fashionable in Europe to bemoan American pop culture as a prime example of the "emptiness" of Western materialism. Say what you like about these Taliban chappies but maybe there's something to be said for beheading anyone caught listening to Herman's Hermits on Kandahar Supergold. As far as one can tell, there are no Saudi Doobie Brothers or Afghan Jill St Johns.

Tempting as it is to regard this as evidence of their superior culture, we should resist. David and Liza's nuptials were not Charles and Diana, but a karaoke royal wedding. Jacko and Liz Taylor are the American equivalents of Mad King Ludwig of Bavaria or those 19th century Ottoman Sultans who were too crazy to be left on their own. The modern celebrity travels with a bigger retinue than your old-time sultan or grand duke. The average rock star is now more hung up on protocol than any count or marquess: A couple of years back, after hosting a grand dinner at his stately home, Sting was forced to issue a public apology for committing a ghastly error in placement and accidentally seating Jools Holland of the band Squeeze next to some no-name session player.

But the crucial difference is that, though Richard Gere may issue tiresome statements on world peace and Bono turns up Zelig-like in the company of a new world leader every other day, in the end, unlike Mad King Ludwig, they have no power over us. It is the genius of America to have disestablished not just the church but the aristocracy. Instead of the latter, we have movie marquesses and DVD dukes and video viscounts moving through the ersatz rituals of their class purely for the amusement of the masses. What an inspired notion. And, by turning her wedding into a grand convocation of the weird, washed-up and wrinkled, Liza only emphasized the benefits: For one day even the hippest hip-hopper will be no more than Little Anthony, with or without his Imperials.

One of the great advantages of a celebrity culture is the way it siphons off so many of the narcissistic and dysfunctional into areas where they can do the least societal damage. Occasionally, the system goes awry and one of them winds up in a serious job (William Jefferson Clinton), but generally things work pretty well. One cannot say the same of Saudi Arabia, whose 7,000 princes are en masse at least as risible and in many cases more tastelessly accessorized than Liza's guests. But their subjects are obliged to pretend they're useful and intelligent: If they laugh at them, they'll wind up laughing their heads off. Likewise, Iraq, where the only celebrity author and musical-comedy star is Saddam Hussein himself: His romantic allegorical novel, *Zabibah And The King,* got great reviews – there's a surprise - and has been turned into a lavish stage production, which is doing sell-out business – there's another surprise.

The tragedy of Iraq is that in order to make it big in showbiz, Saddam had to make it big in mass murder first. Under the American system, his book would have been picked by Oprah, he'd have sold the Broadway rights to Liza's husband, and they'd have signed Petula

Clark and Mickey Rooney for the title roles. No matter how you look at it, that's a massively superior system.

BROWS

High, low, no

June 1996
The New Criterion

I BEGAN READING Michael Kammen's biography of Gilbert Seldes after breakfast. Before breakfast, I'd read my morning newspaper, the Montreal *Gazette*, a sober broadsheet which has just started a weekly rock-video column - and nothing as humdrum and utilitarian as reviews of new rock videos, but rather a provocative and ongoing debate on issues arising and trends discernible therefrom.

The connection between Seldes and *The Gazette*'s latest signing is made explicit in Kammen's cumbersome title: *The Lively Arts: Gilbert Seldes And The Transformation Of Cultural Criticism In The United States.* Perhaps we would have got here one way or another, but it's still doubtful whether, without Seldes' pioneering efforts, 23-year-old female college graduates could have expected (or would have wanted) to make a living commenting on the intellectual themes arising from rock videos.

Sorry, I should have said "music videos." Until fairly recently, "Music" in respectable journals used to mean fellows like Mozart and Beethoven. In *The Gazette*'s and most other arts pages now, "Music" means the Dead Presidents and Niggaz With Attitude, and Wolfgang and Ludwig have been reclassified as the specialist sub-genre "Classical Music." Before Seldes was born (1893), a distinguished music critic was someone who had studied sonata form and soprano passagio. Since Seldes' death (1970), a distinguished music critic has been someone who has studied the social significance of rap lyrics and the postmodern attitudinal irony of grunge haircuts, and can tell you which recording studio Tupac's producer was shot dead outside of.

Seldes would, probably, have disclaimed credit for such a legacy, and certainly it has not brought him fame: I'm sure *The Gazette*'s lady video opinion-former has never heard of him—as far as I can tell, she doesn't seem to have heard of anything before 1981, which is perhaps not such a disadvantage in her new post. But this, too, is part of Seldes' bequest. Before the Great War, criticism was vertical: that's to say, critics specialized and pontificated on literature or music or fine art in terms of the traditions from which they arose. Seldes' generation made criticism horizontal, inventing the concept of the "cultural critic"—the fellow who, today, roams the cultural landscape and is capable of discussing Quentin Tarantino in terms of the Cranberries' new album or the season finale of "Friends" in terms of the seminal Mountain Dew commercial.

That would have depressed Seldes. A Harvard man, he served as managing editor of *The Dial*, and, in 1922, introduced America to "The Wasteland." His most influential book—indeed, his professional epiphany—was a collection of essays from 1924 arguing the merits of vaudeville, Tin Pan Alley, comic strips, musical comedy, and slapstick movies, and its title, *The Seven Lively Arts*, became a kind of brand leader for Seldes, serving in one variation or the other as the label for his columns in *Esquire* and *Park East* and his talks on WNEW.

But the book which followed *Lively Arts*, *The Stammering Century* (1928), was devoted to his second great enthusiasm, "the necessity of historical knowledge and perspective." In the rush to embrace one of his obsessions, we've trampled the other. "Perspective" is now such a debased concept in cultural commentary that the editors of *The New York Times Magazine*, in a special issue celebrating their centenary, didn't seem to notice that they'd actually produced a celebration only of the last forty years. As for "historical knowledge," that crops up only in the most bogus way. I once interviewed Bob Fosse about his films and choreography, and, at some point, I mentioned something about the influence of *commedia dell'arte*. "Congratulations," said Fosse, looking at his watch. "You held out 23 minutes. Most interviewers get to *commedia dell'arte* within the first ten." It was years later that I discovered it was Seldes who first hit upon *commedia dell'arte* as the all-purpose historical precedent: over his long career, he applied the term to Chaplin and the Keystone Kops, Jerome Kern and Irving Berlin, vaudeville and silent movies, and TV sitcoms.

To be fair, Seldes also adapted a famous *commedia* piece—Carlo Gozzi's *The Love Of Three Oranges*—for the Harvard Dramatic Club in 1926. But, for the most part, he was pioneering what became the standard approach in pop-culture criticism: the need to find an intellectual safety net for what regular folks just do without thinking.

Most of us like the pop songs and movies and comic strips of our generation, and the ones that mean most do so usually for non-artistic reasons: it's the film we saw on our first date, the song we danced to at our wedding. Seldes was no different from anybody else in his tastes: as a young man in the Twenties, he liked Al Jolson and Fanny Brice and George Herriman's Krazy Kat comic strip; as an old man in the Sixties, he had less (if anything) to say about Bob Dylan or Lenny Bruce or *The Incredible Hulk*.

I would have liked to have known his thoughts, in 1970, on what had become of his "lively arts," but Kammen doesn't tell us. For the last few years, there's been an increasing trickle of think-pieces by today's graying baby-boom rockers either bemoaning the wretchedness of today's pop music or expressing befuddlement at the way the kids seem to find Hendrix and Dylan and the Stones as boring as the infant rock critics once found Perry Como and Guy Lombardo. But, as a prototype pop-culture warrior, Seldes' own enthusiasms are instructive: They remain generational. No matter how persuasive his analysis, Fanny Brice as Baby Snooks would not seem funny if NBC ran it instead of "Seinfeld" this week. Kammen himself quotes hardly any of Seldes' insights into individual practitioners, an unspoken recognition of a forlorn truth: some of his favorites have endured (the Gershwins, Rodgers and Hart) while many others (Joe Cook, Mutt and Jeff) have not, but, in either category, their fate owes little to Seldes.

Moreover, the best evidence of Seldes' irrelevance to their fate is his own biographer, for, after immersing himself in his subject's entire oeuvre, Kammen has emerged with not the slightest interest in any of Seldes' passions: "Throughout the war, Seldes prepared the program notes, usually for jazz concerts," he writes, the war in question being the one that began in 1939. "He wrote with special appreciation and pride for performers like Bix Beiderbecke."

Bix died in 1931. Whatever Seldes' "special appreciation", it wasn't special enough to communicate itself to his biographer.

Kammen doesn't quote it here, but there's a fine essay Seldes wrote for *Theatre Magazine* in 1924, prompted by the reunion on Broadway of Bolton, Wodehouse, and Kern, which gets to the nub of the matter:

> *It is a pity that you cannot explain or justify delight. Conversation would be so much more amicable if you could. As it is, the friend of your heart or the wife of your bosom, who seems to agree with you on every significant thing in the world, suddenly announces that she cannot abide the Gilbert and Sullivan operettas, and the dead silence that follows indicates that divorce is rapidly setting in.*

Pop culture exists to delight us, and that's the only justification it needs. If it fails to delight, no amount of justification or explanation will commend it to us. It's worth noting that, for all the pressure on the public from an ever proliferating army of zeitgeistmeisters, popular taste seems to have evolved hardly at all: at the end of the century, the biggest-selling pop records (Whitney Houston's devotional ballad, "I Will Always Love You"), the bestselling airport novels (*The Bridges Of Madison County*), the highest-grossing motion pictures (*Sleepless In Seattle*) differ only in the details from their turn-of-the-century predecessors, from "After the Ball" or *The Merry Widow* or the novels of Gene Stratton-Porter. Left to its own devices, the public prefers much the same sentimental romantic fantasy in substantially the same forms as it always has. Each week, the editors of the Arts and Leisure supplements pose the question: What's new? And the truthful answer is: not much.

If the Seldes school of cultural criticism has had a negligible impact on popular taste, what then is it for? When Van Wyck Brooks initiated the highbrow/lowbrow debate in various essays beginning in 1915, the various levels of brows under discussion were the public's: mail-order book companies took up the designations with advertisements demanding to know, "Are we a nation of low-brows?" It's revealing, though, how quickly these definitions came to apply not to the audience but to the art itself. As the old arguments are thrashed out by Kammen, and brow levels fluctuate from high- to low- to middle- from page to page, I found myself recalling Irving Caesar's splendid middle-section to his 1926 hit "Crazy Rhythm":

> *When a highbrow*
> *Meets a lowbrow*
> *Walking along Broadway*
> *Soon the highbrow*
> *He got no brow*
> *Ain't it a shame*
> *And you're to blame.*

There's the history of American art in eight bars: when high art meets low art, soon the high art ain't got no art; when Roy Lichtenstein meets a comic book action hero, the comic book isn't diminished by the encounter, but the "artist" is. "The real problem," noted Louis Kronenberger at a 1952 *Partisan Review* symposium, "is how to avoid contamination without avoiding contact." But, in transforming cultural criticism, Seldes also wound up transforming our cultural vocabulary: what passes for high art now is invariably discussed in terms of its points of contact with pop culture; seduced by those crazy rhythms, the highbrows have become culture's limbo dancers, ever more desperate to please.

Perhaps it was inevitable. At that same symposium, Joseph Frank pointed out that the "European intellectual minority has no sense of guilt at not being part of mass culture." In a monarchical society, it's easier to maintain a hierarchical culture. And, in an age when policy positions on everything from the gasoline tax to abortion wait until the pollsters have crunched the relevant numbers, it's difficult to argue why culture should be exempt: in such a world, why shouldn't the best music or the best book be simply what's Number One?

But brow-wise that's not quite good enough for the limbo dancers of cultural criticism. In his book *The Great Audience* (1951), Seldes suggested that American "folk and popular art" were distinguished by the fact that "both were readily comprehended, romantic, patriotic, conventionally moral." Does that sound like the kind of popular art you read about in the newspapers?

Seldes wrote about comedians as comedians, composers as composers. But most of his successors have lacked either the patience or the technical knowledge to do that. In their place, they've offered a discussion of pop art in terms of its social relevance—or, more accurately, its anti-social relevance. And, year by year, that practice has spread remorselessly so that it now, in Kronenberger's apt word, contaminates our understanding of high culture, too. "In a very real sense," writes Kammen, beginning with a sloppy, meaningless phrase of which he's inordinately fond, "[Seldes] can properly be regarded as a progenitor of the discipline known today as cultural studies." But cultural studies is not a discipline, only a modish means of applying the same trite social observations divined from gangsta rap to Shakespeare and Conrad and anything else that takes your fancy. Once again, Irving Caesar has been proved right: the highbrows ain't got no brows. All the brainiest people in this story—the Ivy League colleges, the leading critics, the arts establishment—reached across to the wilder shores of popular culture and brought back only what was most rotten.

The resilience of the lowbrows, in withstanding our society's cultural leaders, is remarkable and admirable. In a very real sense, as Kammen would say, the casualties of the process Seldes set in motion include both the discipline of criticism and American high art itself (admittedly a delicate flower).

But the third casualty seems set to be pop culture itself. As the popular arts are studied and analyzed ever more avidly, they wind up chasing each other round in circles. The cult stud of the cultural studies circuit is Quentin Tarantino, whose *Pulp Fiction* is endlessly dissected in colleges all over the country. But the film is itself a product of cultural studies: it's a compendium of allusions to a zillion other pop-cultural artifacts. In *Vanity Fair*, in 1922, Seldes wrote that vaudeville was "the only genre I know which can live by burlesquing itself." It couldn't, though. Vaudeville died, as did slapstick, revue, and musical comedy. Today, Quentin and his hordes of Taranteenyboppers can only make movies that look like other movies, pop records that sound like other pop records. Struggling to crawl out from under the incessant bombardment of "cultural studies," the lively arts aren't lively anymore.

EYELIDS

The butler flutters

November 9th 2002
The Daily Telegraph

I Had 60 Pills And A Bottle Of Water In A Lay-By Off The A41. I Was About To Join My Princess... - The Mirror

Queen Blasts Barmy Butler – The Sun

EXCLUSIVE! The butler everyone's talking about opens his heart to the *Telegraph*!

MY STORY by Paul Burble, the last butler to the Conservative Party in those final desperate months before it cracked up, threw away its seat belt and flipped over.

I, PAUL BURBLE, entered the service of the Rt Hon Iain Duncan Smith at the time he became leader of the Conservative Party and I remained by his side all the way to the bitter end, four days later. Sobbing in a dignified manner, I stood vigil over the coffin, with which Central Office thoughtfully provides each incoming leader.

I'll never forget our first meeting. He'd just taken over, and there was an instant bond between us. "This is a very important position," said Mr Duncan Smith. "It requires someone of impeccable discretion who understands that, if you're doing the job properly, people are barely aware of your presence, you're almost invisible, part of the wallpaper."

"Yes, but enough about your job," I said. "What do you expect from a butler?"

As his closest confidant, I helped the young master cope with the tragic break-up of his party and his increasingly frantic attempts to sustain relationships with men. There was the flamboyant Eurotrash playboy Ken, the exotic Mediterranean heart specialist Michael…

"Ken's been in the bathroom an awfully long time," said the master. "Do you think he's a cocaine addict?" But, after crawling through the ventilator shaft and pressing my ear up to the grille, I was able to reassure the poor distraught creature that Ken was merely giving a long phone interview to Radio Berwick on whether the master would last the week.

With Michael it was different. He was a young dynamic surgeon famous as the first Thatcherite to open himself up and install a new heart. One night the master happened to drop by the hospital as Michael was performing a particularly intricate operation.

"Scalpel," he said.

"Scalpel."

"Clamp."

"Clamp."

"The *Liza On Broadway* double CD."

"What's going on?" the master asked.

"It's just a routine procedure," explained Dr Michael. "We're operating on Norman to give him the gay gene."

"Get off me, you dago poof!" yelled the patient, but fortunately the anaesthetist was already increasing the dosage.

I could see the master was intrigued, but how far was he prepared to go? Night after night, I was called upon to smuggle Michael, Ken and dozens of other young men out in the boot of my car to the BBC. "Doesn't anybody want to be smuggled in?" pined the master, tucked up in bed in his flannelette nightgown. "I'm so lonely. Theresa says I need to meet more lesbians and ethnic chappies."

He fluttered his eyelids. I fluttered my eyelids. Unfortunately, we were standing too close and they locked. As we waited for the Special Branch officer to come to our assistance, the master seemed to sense the intimacy of our relationship. "You know I'm the only one who really cares about the Rock," he murmured.

"I know you do," I said, taking his hand, "and I'm very touched. I'll always be here for you. I'm going nowhere. Just like you."

"I was talking about Gibraltar," he muttered brusquely, pulling his hand away. There was an awkward pause. "Would you mind giving me a hand with my urine sample?" he asked.

Afterwards, he made me drive him around the back streets of Paddington, handing out £50 notes to prostitutes. He'd wind down the window and say, "Here's the Tory Party's new Unmarried Hooker's Winter Coat Allowance." They'd tuck it down their cleavage, and say, "'Ere, fanks very much, Jeffrey."

"Actually, it's Iain," he'd say. "All part of our commitment to community outreach in the new Britain."

I'll never forget one amazing evening when he came downstairs looking absolutely stunning in a fur coat and diamond ear stud. I drove him to the Brompton Hospital and he stepped out the car and dropped his coat, and underneath he was completely naked! He stood there for a while but none of the passers-by seemed to notice him and after about 40 minutes his teeth were beginning to chatter. "T-T-Theresa's idea," he said, getting back in the car.

The following week, I had an audience of Mrs May. I remained standing throughout our three-hour conversation, as she'd sold off the chairs as a cost-cutting measure. The master's beaker was on the desk. "It's come back positive again," she said.

"Sorry to hear that, Ma'am," I said.

She passed it over to me and read out the lab report. "97.4% heterosexual white male urine, 1.1% Earl Grey, 0.6% Marmite, 0.5% Horlicks, 0.4% Cooper's Thick-Cut Oxford Marmalade." The mistress looked at me sternly. "You know these are all on the Tory Party's new list of prohibited substances," she said, shaking her head sadly. "I've tried to reach out to Iain so many times."

"The trouble is, Ma'am," I said, "that you speak in colour. Mr Duncan Smith speaks in black and white. Er, I mean..."

"What's all this about a gay scandal below stairs, Burble?"

"Oh, don't worry about that, Ma'am," I said. "Just some obscure bill. It'll all be hushed up. Shouldn't even make the papers, I expect."

EYES

Not seeing is believing

November 17th 2002
The Chicago Sun-Times

I T'S AN OCCUPATIONAL hazard of media analysis that any two pundits can scan the horizon and discern entirely different political landscapes on the road ahead. Usually, of course, our predictions conform to our particular biases. But it's slightly more alarming when the analysts can't take their blinkers off *after* the event. They're looking at what's happened, but they don't see it.

Say you're a newspaper editor trying to figure out what big analysis piece to splash over your front page the Sunday after the election. What would you go with? Well, *The Valley News*, the biggest daily in the western part of New Hampshire, surveyed the scene the morning after: A state that had been high up on the list of Democratic targets had instead voted for an all-Republican congressional delegation, an all-Republican executive council, a Republican governor, and a 75 percent Republican state Senate and general court. Nevertheless, the editors dutifully looked up "Post-Election Analysis Features You Can't Go Wrong With No Matter How Many Times You Trot Them Out" in the Columbia Journalism School Book of Lame-O Cliché Stories You Can Serve Straight From the Freezer and turned in a somnolent front-pager headlined "Women Candidates Fared Poorly In Midterm Elections."

As it happens, certain women candidates fared rather well in the elections: Elizabeth Dole and Katharine Harris, for example. On the other hand, the Widow Carnahan in Missouri and Kathleen KENNEDY!!! Townsend flopped out. I wonder what could be the reason for the remarkable disparity in how these women fared. Might it have something to do with the fact that the former are Republican women and the latter Democrat? Perish the thought! Pondering the fate of Mrs Carnahan and Jeanne Shaheen, *The Valley News* and its interviewees - spokeswomen for the National Organization for Women, etc - thought that women's issues such as "reproductive rights" had been overshadowed by the way Bush had gone around whipping up a lot of "fear" about obscure fringe issues like national security.

If you were really interested in doing a story on women and the elections, it would be this: "Women Candidates Backed By So-Called Women's Groups Fared Poorly." The National Organization for Women endorsed 18 candidates in this election: 15 lost. When the women anointed by the abortion absolutists at NOW, NARAL and Emily's List bomb that badly, it should suggest even to our dopey press that perhaps the rent-a-quote spokeswomen don't represent quite as many women as they claim to.

The other story that might be worth going with is "Young Women Hot For Republicans." Indications are that, in this month's election, the famous "gender gap" from which the GOP's country-club old-boys executive-men's-room sexists are always said to suffer was wiped out among young voters. To be honest, I've never really subscribed to the "gender gap" theory. After all, a gender gap cuts both ways, and in recent years the Democrats have arguably suffered more from their lack of appeal to men. But on November 5th, guess what? Among female voters under 30, as many voted Republican as Democrat. The Dems are the party of old women. Oh, okay, "mature" women.

But come on, does anyone honestly vote like this? If I've got a choice between Condi Rice and Ted Kennedy, I'll go with the broad. If it's Don Rumsfeld vs Nancy Pelosi, I'll vote my gender. And, believe it or not, most feminists do the same thing: If it was Elizabeth Dole vs Bill Clinton, the need to elect women would take a back seat to the need to elect a "pro-choice" serial pants-dropper. The only people who think in these terms are folks like Judy Woodruff, who late in the evening on CNN, with Democrat hopes crumbling to north, south, east and west, suddenly decided that there was a pressing need to discover how "women" were doing in this election and commanded the back-room psephologists to unearth the relevant data. Now, scarce a week later, Judy is again all a-twitter because Nancy Pelosi has become the first woman in Congress to be elected party leader.

Who cares? Just about the least interesting thing about Mrs Pelosi is that she's a woman. What's interesting is that she's a Haight-Ashbury leftist who voted against war with Iraq. That's likely to prove more relevant in the two years ahead than whether she looks better in a bikini than Walter Mondale. What is a "women's issue" anyway? To some, it might be the sacred constitutional right to avail oneself of a partial-birth abortion. But to others it might be the war on terror. After all, if there's one single issue that distinguishes Western values from Islamofascism, it's the treatment of women. Imagine being forbidden by law to go to school or leave the house unaccompanied. Imagine the state deciding what clothes you can wear. Imagine being prevented by the government from ever feeling sunlight on your face. I'd say voting for people who liberate women from theocratic fascism is a women's issue.

Most American voters aren't interested in candidates because they're women, or because they're widows, or because they're triple-amputees, or because they're last-minute iconic replacements for suddenly deceased senators. Believe it or not, right now they're interested in a couple of overriding issues. A not insignificant segment of the electorate has moved in one direction and, if the Democrats aren't to do worse next time (the Senate seats they'll be defending give no cause for optimism), they have to figure out a way to get that segment to move back toward them. Who is this segment? And why does it prefer the Republicans? Some of us reckon we know. But the media keep yakking on about "women's issues" and all the other pre-9/11 trivia as if all the king's clichés and all the king's bumper stickers can put Humpty Dumpty Democrat together again.

"Don't Stop Thinking About Tomorrow," sang Fleetwood Mac in Bill Clinton's '92 campaign. Ten years on, the Dems can't stop thinking about yesterday. For Al Gore, it's always Florida 2000 and his chads are dangling. For the National Organization for Women, it's always 1973 and Roe vs Wade. For Jesse Jackson, it's always 1963 and Selma, Alabam'. Come election time, the Democrats sound like an oldies station with only three records.

Even in crude terms of "shoring up their base," the Dems feel clapped out: When Harry Belafonte has nothing to say about America's Secretary of State and National Security Adviser other than that massa's been roun' the cottonfields and picked out a couple new house slaves, he's cranking out a tired refrain even older that his last hit. When elderly feminists run around warning "women" that their right to an abortion is at stake in this election, it's not just that it's untrue, it's that it's so lame, and frankly sounds just plain loopy when North Korea has nukes and there's a Homeland Security bill to pass. We live in interesting times, as the old Chinese curse has it, and the Dems have nothing interesting to say. In the midst of a great historical drama, they're still doing vaudevillian knockabout. On all the important issues, the Dems were profoundly unserious. And their complaint that the President's selfish obsession with national security, foreign policy and other trivialities was crowding out "the real issues" is a good example: Whatever the merits or otherwise of a "prescription drug plan for seniors," a great national party has to be more than a pharmacist-in-chief. The obsession with pill dispensing sounded weird in 2000, and just plain inadequate now. Late in the evening, as the band at Missouri campaign HQ attempted to rouse the despondent crowd with Steve Allen's great song "This Could Be The Start Of Something Big," you couldn't help feeling that the Dems face the opposite problem: This could be the end of something small.

It's not just the Democrats who'd benefit from a little self-examination. What about all those network boobs who gave us the fawning puff pieces about how Bill Clinton's crowd-pulling rock-star charisma is bigger than ever on the campaign trail? Sure, he pulls crowds - of Republicans, to the polls. You can pretty much correlate the Democrats' worst results on Tuesday with Bill's travel schedule during the campaign. The polls had Bill McBride holding his own against Florida Governor Jeb Bush until Bill and Al Gore showed up to lend a hand. The same in Maryland, where Kathleen Kennedy Townsend was still the favorite until Al and Bill breezed into town to stump for her.

And, as bad as the Wellstone "memorial service" was, it wasn't as lousy as the media coverage of it. On the following morning, CNN's Jonathan Karl reported that "the overflow crowd came as much to celebrate Paul Wellstone's life as to mourn his death," and referred only to the "impassioned appeal" made by the senator's son. The boos for Trent Lott? The walkout by Governor Jesse Ventura? The totalitarian hectoring by Wellstone aide Rick Kahn as he singled out attending Republicans by name and demanded they switch sides? Jonathan Karl sat through all of it and evidently thought none of it worth mentioning. It was the same with Jodi Wilgoren in *The New York Times*, whose report of the event – "Mourning In Minnesota" - seemed blithely unaware of its tenor. Kahn's partisan bullying was described only as a "spirited eulogy." Karl and Wilgoren missed the story: They saw what millions of American TV viewers saw, but they were either blind or averted their eyes.

Remind me never to complain about "liberal media bias" again. Right now, liberal media bias is conspiring to assist the Democrats to sleepwalk over the cliff.

CHEEK

Dimpled

November 17th 2000
The Daily Telegraph

WELCOME BACK to Campaign 2000 Election Update: America's Day Of Indecision - Day 11. The show ain't over till the Four Chads have sung, and few vocal groups are as popular with Democrats as this quartet.

That's mainly because, just when you think they've got to the big finish, someone goes, "One more time!" and back they go to the beginning. That's why starstruck Gore lawyers keep posters of their favourite Chad pinned up on their bedroom walls. Who's your favourite? Dimpled Chad? Swinging Chad? Pregnant Chad? Hanging Chad? Hey, don't forget Flat Chad, though, for obvious reasons, he doesn't get to sing with the group - so far, anyway.

If you want to catch the Chads for yourself, check the Butterfly Ballot – that's not a Florida nightclub but Palm Beach County's now famous voting form, on which the chads are the little bits of paper you punch out next to your preferred candidate's name.

Whether or not he ever becomes President, Mr Gore has done this Republic an important service. Thanks to the Vice-President, we now know that, when you let ballots get counted mechanically, the machines tend to display their extreme Republican bias by failing to discover enough additional votes for the Democrats. Fortunately for the massed ranks of Gore lawyers, the folks counting by hand are able to divine the voting intentions of Floridians too weak to punch out the chad.

"There's enough separation to consider it a vote for ...Al Gore," said David Leahy, Dade County elections supervisor, holding the ballot up to the light. On Tuesday, the Republicans urged the county to adopt a uniform standard for what constitutes a valid vote, but the Canvassing Board would have none of it. "I won't know it until I see it," said county judge Lawrence King.

Fortunately, most Democratic officials can be relied on to see it remarkably often. For example, if you see a Hanging Chad – that's a chad clinging to the ballot by just one corner – that's most likely a vote for Gore. So might a Swinging Chad - a chad affixed by two corners.

However, let's not forget the Pregnant Chad, a chad still firmly in the ballot but with a visible bulge, and the Dimpled Chad, also firmly in but with a vaguely discernible indentation.

Previously in Florida hand counts, the Pregnant and Dimpled Chads have not been regarded as proper grounds for a recorded vote. Still, the Palm Beach Election Canvassing Commission, after relaxing its criteria a quarter of the way through the first manual recount, had been discovering legions of hitherto unknown Hangers and Swingers for Gore.

Observers at the count report that, by the time the ballots have been manhandled by Democrat Carol Roberts and her fellow supervisors, some hitherto Swinging Chads have become Hanging Chads and some Pregnant Chads have become Swingers.

Even these generous interpretations failed to uncover enough Gore voters to put him in the White House. So on Tuesday Democrats sued to get Pregnant and Dimpled Chads included as votes for Gore, and on Wednesday, for the first time in Florida history, Judge Jorge Labarga ruled that Preggers and Dimply could be counted. Of course, in holding up a ballot to determine its status, you might take a perfectly smooth Flat Chad, one with no discernible indentation, and accidentally Dimple it.

Whether even this would produce enough votes for Gore is unclear. The courts seem to have accepted the principle that Democrats have particular difficulty voting, and that not to allow party officials to identify voter intent retrospectively is, ipso facto, discriminatory.

According to Jack Davis, "California's premier expert on hand counts", "The voting profile shows clearly that Democrats are more likely to screw up the chad than Republicans. This goes across the board by at least one per cent."

So, if the Bush team objects to hand counts, it might be easiest just to increase Gore's vote in any chad-minded county by one per cent. If that's not enough, we need seriously to consider whether the Flat Chads - ballots with no mark on them whatsoever - are not, in fact, Gore votes. After all, there may be thousands of voters in Palm Beach too feeble-minded even to be aware that they are actually Gore supporters.

Come to that, it seems likely that the millions of Floridians who didn't vote at all are also Gore supporters who were just delayed at the shuffleboard court. Why should they be disfranchised? No doubt that's why Democratic State Representative Irving Slosberg was found to have a Votamatic ballot-punching machine in his car. "I asked Mr Slosberg to return it to me," said Denise Cote, Director of Public Affairs, "and he said no, he intended to use it." Way to go, Irv!

Thanks to the likes of Irv and in an admirable spirit of diversity and multiculturalism, Al still has a shot at becoming the first Leader of the Free World from a banana republic. We are not in Florida or Kansas anymore: We are in Chad. Al's come out Swinging; Dubya is Hanging at his ranch; Katherine Harris, Florida's Secretary of State, is cutely Dimpled in a steely sort of way; and the air is Pregnant with excitement.

On Wednesday, Mrs Harris told the errant hand-counting Gore counties to take their chads and shove 'em, and certified the votes as they stood. Gore's job now is simply to tie the certification up in court so that, when the Electoral College votes in December, Florida won't be represented and he'll win a majority of those there. As the pressure to produce a valid ballot has gradually lessened - from Hanging to Dimpled - so the pressure on the electoral process has been ratcheted up.

Gore, if he wants, can now tie the certification up in court interminably, forever tainting any shred of legitimacy George W. Bush might still have. Will he do it? What do you think? This post-election face-off has been, according to the pious network anchors, a "civics lesson", but it would be more accurate to call it a class in thuggery. Two years ago, worldly Dems told us not to worry: The corruption was strictly confined to oral sex, and sophisticated chaps that we are, we could all understand that, couldn't we? Now it seems clear it was just a dry run for a more ambitious project.

Even in this week's statement offering "completion", Gore couldn't resist putting a tighter squeeze on Democratic officials. "We should complete hand counts already begun in Palm Beach County, Dade County and Broward County," he declared, not to "count" the votes but rather to "determine the true intentions" of voters.

Yet heavily Democratic Dade voted not to recount, and is now being sued by Gore to force it to do so. Thus, the ever more pressured condition of Dade's Democratic officials: First, they were just being lightly Dimpled by the Gore heavies; now they're Hanging by a thread. Florida voters may not know how to punch properly, but Al and his lawyers do.

NOSE

Dive

March 11th 2001
The Sunday Telegraph

MICHAEL JACKSON flew into disease-ridden Britain last week, wisely taking the precaution of wearing a surgical mask. At Blenheim Palace, which must have seemed a bit of a comedown after Neverland, they rolled out a red carpet covered in disinfectant. Just for a change, instead of grabbing his crotch, he was grabbing his crutch, hobbling around due to some domestic misfortune. Speaking at the Oxford Union, he called on the world to adopt his Children's Bill of Rights, including "the right to be thought adorable" and "the right to be listened to without having to be interesting". The right to a $30 million out-of-court settlement from famous wealthy playmates was not mentioned.

Michael also revealed the pain of his own lost childhood, as tears rolled down his cheek - or whoever's cheek it was originally. It is a constant motif in his work. "Have you seen my childhood?" he sang in "Childhood", the theme from *Free Willy*. (*Free Willy*, by the way, is a motion picture and not another demand from his Bill of Rights.) The artwork of his 1995 double-album *HIStory, Past, Present & Future - Book 1* includes a self-portrait of Michael as a young boy clutching a microphone and huddled in a corner:

> *Before you judge me*
> *Try hard to love me*
> *Look within your heart and ask*
> *Have You Seen My Childhood?*

Michael spent his childhood pretending to be grown-up enough to sing love songs with the Jackson Five. He's spent his adulthood pretending to be a child. For a while, he liked to hang out at Disneyland with Mickey Mouse, one of the few A-list celebrities with whom he had anything in common - not least the white gloves, squeaky voice, snub nose, bizarre albino face bearing no relation to the jet black surround, and a penchant for hanging out with kids even though you're well into middle age. Later, he was friends with *Home Alone* cutie Macaulay Culkin: they liked to go shopping together wearing buck teeth and false noses. But Macko outgrew Jacko and moved on to broads and booze. The last time Michael was in Britain he was accompanied by Omar Bhatis, a 12-year old boy who came first in a Michael Jackson lookalike contest in Norway: kitted out in matching white gloves and surgical masks, they checked into the Dorchester together.

For his latest visit, he was accompanied by a new best friend, Rabbi Shmuley Boteach, who, unusually for a Jacko chum, appears to be old enough to shave. His other friend is Uri Geller, for whose renewal of marital vows Michael served as best man. Mr Geller is best known for his ability to take a spoon and, by all but imperceptibly gliding his finger over the surface, bend it into a different shape entirely. You can also do that with Michael's nose, of course, but he tends to get annoyed. Still, the sight of Michael Jackson in the company of men old enough to be the fathers of his previous friends prompts a question: is the Peter Pan of pop finally growing up?

Born in Gary, Indiana in 1958, young Michael enjoyed eight relatively showbiz-free years before being forced with his older brothers into a singing quintet. According to the authorised biopic *The Jacksons: An American Dream*, his ambitious father isolated Michael from other boys, so that his only company was a small rodent who scurried around the kitchen floor. "Will you be my friend, Mister Rat?" he asked, and Mister Rat, no doubt cynically contemplating the prospect of a massive sexual harassment suit down the line, twitched his nose in agreement.

But, a few nights later, the Jackson Five won the high school talent competition, and Michael came home eager to show off his trophy, only to find his little chum under the kitchen table, dead in a rat trap. "He was my friend. Somebody killed him!" wailed the young singer, racing around the house accusing members of his family and trembling on the brink of inverting the old Jimmy Cagney line: "You dirty brother, you killed my rat!"

At Tamla Motown, the Jackson Five's first four records went to Number One and one of them, "I Want You Back", with an 11-year-old Michael wailing about an intensity of passion he knew nothing about then and seems unlikely ever to experience, is as good as anything Motown ever released. He didn't think anybody would want him without his brothers, but *Thriller* (1983), produced by Quincy Jones, artfully fused soul, rock, Vincent Price and the nascent video form to become the world's all-time best-selling album. Yet, though still just about recognisably African-American, Michael was already considerably less black than on the cover of his previous album, *Off The Wall*. He was also rumoured to be hanging out with chimps and llamas.

In an in-depth interview with Oprah Winfrey, the King of Pop mused on the preoccupations of the press. "If I had a chance to talk to Michelangelo," he squeaked, "I would want to know about the anatomy of his craftsmanship, not about who he went out with. That's what's important to me."

"How much plastic surgery have you had?" responded Oprah, less interested in the anatomy of his craftsmanship than the craftsmanship of his anatomy.

"You can count it on two fingers," replied Jacko, holding up two he'd made earlier.

But, as with Michelangelo's David, Oprah's eye was drawn to one region in particular. "Why do you always grab your crotch?" she asked, alluding to his principal choreographic innovation.

"It happens subliminally," he said, although a more plausible explanation is that he's just checking on the one bit of him the plastic surgeon hasn't got to.

There are those who say Michael changed after his hair caught fire while filming a Pepsi commercial. There were stories that he took female hormones to keep his voice high, that he slept in an oxygen chamber, that he lightened his skin. By the Eighties, his celebrity friends were mostly post-menopausal women such as Katharine Hepburn and Sophia Loren. When asked whether he'd proposed to Elizabeth Taylor, his lips remained sealed, though that may be just an unfortunate side-effect. It could be that the marriage story was simply a misunderstanding: he asked Liz for her hand and she said: "Why not? You've already got Diana Ross's nose." By the time he eventually married Lisa-Marie Presley, it seemed more to do with the King of Pop's dynastic ambitions, a desire to mate with the essence of Elvis and sire the greatest pop star of all.

But there was no progeny or even, by some accounts, much heavy petting. By the mid-Nineties, Jacko was in the papers mostly for settling out of court with two boys and prompting California prosecutors to fly to Australia to interview another. Eleven-year old Brett Barnes told investigators he and Jackson had slept in the same bed together but insisted the singer had behaved properly at all times. Reeling from the allegations and hooked on painkillers, Jacko checked into Beechy Colclough's Charter Clinic in Chelsea, where Beechy weaned him off his addiction by substituting tea and biscuits for the tranquillisers. In Jacko's later videos that may look like his crotch he's grabbing, but actually he's just checking on his packet of Hob Nobs.

Since then, Michael has had two children by Debbie Rowe, the nurse who'd been treating him for his alleged pigmentation disorder. Michael's son is called Prince Michael Jr, his daughter is Paris Michael, the name "Princess Michael" evidently being thought likely to expose her to ridicule. The proud father, wearing a surgical mask over his surgical mask, was present at the birth of both children. Whether he was present at the conception is the subject of much speculation.

It's hardly worth mentioning the records any more. After *Thriller*, the follow-up *Bad* was considered a flop: Michael Jackson became a symbol not just of his own weirdness, but of the insanity of an industry where sales of 25 million make you a loser. You could argue that Michael Jackson's ever-more-bleached complexion is a shorthand for pop's history: after a century of white exploitation of black music, he's the first black singer to become his own white cover version. Does Michael ever look in the mirror, recall that little boy with the Jackson Five and think "I Want You Back"?

NOSTRIL

Gushing

September 13th 1998
The Sunday Telegraph

IN CHAPTER five of *It Takes A Village To Raise A Child*, Hillary Rodham Clinton recounts her first attempt at breast-feeding:

As I looked on in horror, Chelsea started to foam at the nose. I thought she was struggling or having convulsions. Frantically, I pushed every buzzer there was to push.

A nurse appeared promptly. She assessed the situation calmly, then, suppressing a smile, said, "It would help if you held her head up a bit, like this." Chelsea was taking in my milk, but because of the awkward way I held her, she was breathing it out her nose!

A similar situation has now befallen the other child in Mrs Clinton's life. Even a cursory glance of the rise of Bill and Hill suggests that most of the awkward stuff, from that long-ago Arkansas land deal to the firing of the White House travel office, was handled by Hillary, the brains of the outfit. And, while one might detect a pattern of unseemliness, Mrs Clinton was far too clever to leave any fingerprints. Unfortunately, the chief exec of Clinton Scandals Inc left her husband in charge of one small branch office: sex. And, sadly, Mr Clinton's small branch is in danger of scuppering the whole operation. As with Chelsea and her nostril, the impenetrable dyke has sprung a leak at its weakest point. Unlike his canny wife, the President has left a trail of fingerprints and more on everything and everyone he touched.

So, in the end, the Starr Report was "just about sex" – or, more precisely, sex and the legal complications arising therefrom. On Friday the President's supporters were handed a list of "talking points" to bring up at interviews: Whatever happened to Whitewater? And Vince Foster? And Travelgate and Filegate? And all the other scandals Ken Starr was supposed to be investigating before, like a floundering Texas wildcatter, he finally struck the rich, deep gusher of presidential semen?

At their press conference on Friday, his attorneys essentially agreed with the facts. What they dispute is what they mean. It doesn't count as perjury because oral sex doesn't count as sexual relations. Fondling Monica's bare breasts doesn't count as sexual relations. Staining her dress doesn't count as sexual relations. Sexual relations doesn't count as sexual relations. For the purposes of law, it counts as sexual relations only if it's with Hillary Rodham Clinton.

The President's defence has now evolved into a modification of the good cop/bad cop routine. The ludicrous hair-splitting and attacks on the independent counsel have been delegated to the attorneys. In Dublin, Mr Clinton said what he did was "indefensible"; in Washington, they're defending it as vigorously as ever: it's private and not impeachable - even though Monica performed her non-sex act on the Oval Office broadloom and while the President was on the phone to three Congressmen. Just as it's his attorneys' position that, while she may have been having sex with him, he wasn't having sex with her, so too the President was acting as Head of State and Commander-in-Chief above the waist but as a private citizen below. For their part, the American people seem increasingly inclined towards the Clinton proposition that "even Presidents have private lives".

Meanwhile, Mr Clinton has embarked upon a strategy of continuously apologising for his previous apology, which it seems, though legally accurate, was not appropriate. So the President is now abasing himself in front of anyone willing to keep a straight face: he is a sinner; he seeks our forgiveness; he vows to repent. Most Americans fondly recall this act from the 1980s, when one big-bucks televangelist after another struggled to recover from the usual loose-lipped hookers: "O, Lord, thank you, sweet Jesus, for placing this lush, nubile temptation in the path of your humble servant to test his faith and bring him through the fires of fornication until he can stand before you, chastened if refreshed, and say Hallelujah, for I am truly saved! Let him who is without sin cast the first stone. The rest of you send cheques made payable to my legal defence fund."

A decade ago they used to say, "Let Reagan be Reagan." On August 17th, for four minutes on national TV, they let Clinton be Clinton - and he's been in freefall ever since. The mask dropped and for most viewers, it was the first glimpse of the real Bill Clinton - arrogant, defiant, self-pitying. On Thursday, in what was reported as a "therapy session" with his Cabinet, the mask slipped again: when Donna Shalala told him she felt betrayed at having been deliberately deceived and then sent out to broadcast the lie, the President rounded on her for her disloyalty in his time of need. (Mr Clinton has held just two Cabinet meetings this year: the first, to lie to them; the second, to apologise for lying to them.)

So many needs, so little time. Mr Clinton is spinning without a script these days, still relying on a legal defence for what is a political problem. No Democratic candidate has anything to gain by defending the President. But every single one of them has nothing to lose by distancing himself - in a principled, non-partisan, country-above-party kind of way. It may, indeed, be to Mr Clinton's advantage to provoke Democrats to speak out against him: that may save just enough of them to prevent the Republicans taking 60 seats in the Senate. And the fewer Republicans there are in the Senate, the better Mr Clinton's chances.

In the short and medium term, things will get worse. But, if he keeps his nerve and he's prepared if necessary to bring the country (like Monica) to its knees, the long-term looks more promising. Presidents go down in history in one line: Ike? Won the war. JFK? Got assassinated. LBJ? Done in by Vietnam. Mr Clinton is destined to be the sex-fiend President. The only thing left to play for is whether he's the sex-fiend President who beat the rap. So the choice now is whether he joins Andrew Johnson in the Impeachment League or Richard Nixon in the Quitters' Club, and, in the pantheon of Presidents,

Johnson comes a notch above Nixon. Clinton won't resign, one congressional Democrat says; they'll have to handcuff him and drag him out with a tractor. It's true that, at this stage in his downfall, Nixon was also resisting resignation. But, unlike 1974, Mr Clinton has Nixon's resignation as a precedent; he's already sick of being called William Milhous Clinton and the only way he can scotch that is by not doing what Nixon did.

Andrew Johnson, by contrast, appeals far more to Clinton's sense of himself. Johnson was impeached in 1868 for unconstitutionally sacking the Secretary of War. However, in the ensuing trial, all sorts of other flimflam was dragged in including allegations that he was implicated in Lincoln's assassination. For Bill Clinton, who believes the vast right-wing conspiracy has been trying to stick him with everything from Colombian drug-running to unsolved Arkansas murders, Johnson's ordeal strikes a chord: he really does feel this guy's pain. More importantly, Johnson won: the Senate vote fell one short of the two-thirds majority required. Within a few years, the attempt at impeachment came to be seen as a mistake and the Johnson Presidency was remembered - albeit barely - for the purchase of Alaska and various other second-rank achievements.

Clinton and Johnson are kindred spirits - aside from the fact that Johnson wasn't a boorish fellatio junkie. But in 1868 he was the Comeback Kid - just like Bill Clinton in Little Rock in 1982, in the snows of New Hampshire in 1992, in Washington after the Government shutdown in 1995.

The saddest moment in the Starr Report comes when poor, foolish Monica tells the grand jury that the President had fallen in love with her. In reality, Bill Clinton is in love with the myth of his greatness: he can quit now or take a shot at the unlikeliest comeback of all. And, as she did with Chelsea and her frothing nostril, Mrs Clinton will be there to help prop up her little boy until the unfortunate dribble of DNA is merely the latest ineffective stain on his record.

Dumb and Dumas

March 21st 1998
The Daily Telegraph

OFFERED YET another costume drama, Broadway impresario Lee Shubert brushed it aside. "Audiences," he said, 70 years ago, "don't want plays where people write letters with feathers." You'd think it would be even truer now, when audiences don't even want plays where people write letters with ballpoint pens. Today, every thriller contains a scene in which the hero has to hack in to the computer and download the secret formula before the security guard returns from the men's room.

Yet, even in our dismal electronic age, the feather writers are holding their own - the scratch of quill, the crackle of parchment, the blob of wax, the royal seal... And the all-time king of the Hollywood feathermen, Alexandre Dumas, is proving to be indestructible. Dumas produced novels that have the streamlined energy of cracking film scripts - which may be why everyone keeps going back to them. In the past couple of years, we've had an authentically French *La Reine Margot* and an appallingly vulgar *Three Musketeers*, in which Athos, Porthos, Aramis and D'Artagnan were effectively reduced to a quartet of Californian beach bozos. This week, it's the return of *The Man In The Iron Mask*, this time with musketeers Gerard Dépardieu, John Malkovich, Jeremy Irons and Gabriel Byrne flouncing around the French court in frilly cuffs and plumed hats and Paula Jones hair.

When publicity still from the film landed on my desk, I assumed the limpid moon-faced beauty framed by brunette tresses was Helena Bonham Carter. On closer inspection, it proved to be Leonardo DiCaprio, who plays both King Louis XIV and the eponymous leadhead. Hollywood's view is that, after *Titanic*, moviegoers would shell out to watch Leo clipping his toenails. Despite indifferent reviews and an incomprehensible trailer, they seem to have been proved right: in America last weekend, *The Man In The Iron Mask* pulled in more than $17 million. All the same, that was still a couple hundred million shy of *Titanic*, suggesting that American audiences on balance prefer Leo's angelic features in a cloth cap rather than encased in metal. "Why's he have to wear an iron mask?" asked the girl behind me, a big Leo fan.

"What do you mean?" I said.

"Well, it's just, like, not him."

"I don't think it's meant to be a fashion statement," I said. But who knows? The greatest musketeer of them all, Douglas Fairbanks Snr, turned D'Artagnan into one of the most influential fashion statements of the decade. His 1921 *Three Musketeers* cost $1 million, which made it at that time the most expensive film ever. Milady's gowns were genuine antique brocade and the stones for the 17th-century buildings were hand-hewn into irregular shapes so they appeared more weathered. Having gone to all that trouble, Fairbanks felt the least he could do was grow a period moustache. His wife, Mary Pickford, and his friends cautioned against it. The moustache, they said, would destroy his career. In the America of 1921, moustaches were worn by fat silent-screen comics, elderly gentlemen and, in the parlance of the day, "powder puffs".

But Fairbanks insisted. And, singlehandedly, he brought the moustache back into fashion. Without Fairbanks, we wouldn't have had Errol Flynn's 'tache in *Robin Hood*, or Ronald Colman's in *The Prisoner Of Zenda*, or maybe even Charles Bronson's in *Death Wish*. In 1921 Fairbanks started a nationwide craze for pencil moustaches among American men. It was the most influential moustache in screen history, and, ever grateful, Fairbanks kept his till the end of his life.

It's safe to say that he would be nonplussed by the new *Iron Mask*. The opening scene features an impressive assessment of the, ah, manhood of Aramis (Jeremy Irons) and a noisy fart from Porthos (Gerard Dépardieu). It's not necessarily inauthentic: no doubt 17th-century musketeers discussed each other's penises and farted extensively. But it's not romantic heroism as Fairbanks would have understood it.

In 1920, Fairbanks acquired the rights to a newspaper story called "The Curse of Capistrano" and turned it into *The Mark Of Zorro*, the masked swordsman with a penchant for carving his monogram on the face and, indeed, the buttocks of evil men. This year, Zorro returns again, in the fetching shape of Antonio Banderas, attempting to leave his distinctive initial somewhere about the persons of wicked colonial grandees. Not much has changed since 1920. The masked man is still Senór Zorro by night and the feebly foppish Don Diego de Varga by day, righting wrongs in the pueblos of Spanish California in the early 19th century. The beautiful Marguerite de la Motte worships Zorro but despises Don Diego, being unaware that they are one and the same. This is tough on the poor guy, who adores Marguerite but dare not reveal his true identity: love means having never to say you're Zorro.

The Mark established Fairbanks as a sex symbol, not because of his stiffness in the love scenes but because of his thrusting elegance with a sword. In the days before explicit sex on screen, Fairbanks made his blade the quintessential projection of manhood, along with his moustache. The film spawned a host of imitations throughout the Twenties and Thirties: In *The Desert Song*, the simpering son of the Governor of Morocco by day is, by night, the mysterious Red Shadow. In *Rio Rita*, a Mexican general is also the mysterious Kinkajou. The plot features a French nobleman in disguise in the French colony of New Orleans in the 1780s. The only problem is that in the 1780s New Orleans was a Spanish colony. On balance, this is probably a more cavalier error than that of the 1979 *Prisoner Of Zenda*, in which can be glimpsed, on the streets of the Ruritanian capital, two Volkswagen Beetles.

Given the amount of money lavished on these remakes, the big threat these days is not inauthentic scenery or costumes but the inauthentic actor inside them. As the most recent *Three Musketeers* made all too plain, a modern sensibility somehow taints and tarnishes the gayest of blades. In such a world, the fate of even as reliable a crowd-pleaser as Dumas is bound to be precarious. But, even if Leo in the iron mask fails to please, even if the new *Zorro* is greeted with a massive zzzzzzz, the swashbuckler will still have left a permanent mark on Hollywood. After playing Zorro, Douglas Fairbanks decided he'd like to live the life of a real Don Diego, so he bought a 3,000-acre Spanish hacienda outside San Diego and renamed it Rancho Zorro. The first thing he did was get a new irrigation dam built. As he and Mary Pickford examined it, they decided, on a whim, to press their hands into the wet cement and inscribe their names. It looked kind of cute. And then they remembered that their friend Sid Grauman was about to open a new movie theatre in Hollywood and was looking for a gimmick. What about handprints and signatures in cement? There today, on the sidewalks of Los Angeles, is the real mark of Zorro.

LIPS

Smoke rings

July 16th 1997
Slate

T HE EARLIEST smoking song I've ever come across is "Tobacco's But an Indian Weed" - from the late 1600s, which seems a bit slow off the mark: Sir Walter Raleigh had brought the first tobacco leaves back from the colonies to Queen Elizabeth almost a century earlier. On the other hand, he also brought back the potato, and how many great potato songs had anybody written by then? But sooner or later everything winds up in the ashtray of history and, 300 years after that first entry, it seems almost certain that the Cigarette Songbook has no new leaves to turn over. For that reason alone, k d lang's *Drag* will prove a useful anthropological document, rounding up as it does some of the smokiest songs of the century, from "Smoke Rings," the old theme of Glen Gray and the Casa Loma Orchestra from the 1930s, to Steve Miller's Seventies rocker "The Joker" ("I'm a smoker, I'm a midnight toker").

The first thing to be said is that Miss lang - or maybe it's miss lang - lives up to her title: *Drag* is a deep, languorous inhalation. Its orchestrations, especially David Tom's guitar loops, are as near as anyone's ever come to the sensation of smoking, even if by the time she gets halfway through the Hollies' "The Air That I Breathe" – "Peace came upon me/ And it lee-ee-ee-eaves me weak" - she seems to be unwinding from a heavy night at the opium den rather than down to her last Marlboro Light.

The second thing to be said is that *Drag* is a pun: On the cover, lang is wearing pinstripes and a ruby cravat, like Oscar Wilde heading out clubbing. It's as much about sexual role play, and smoking as a metaphor for love, every popular singer's real addiction. (If any male vocalist is looking for an equally adroit album title encompassing both cigarettes and sexuality, may I suggest, at least for those versed in the divergences of British and American slang, *Fag*.) Wilde, in *The Picture Of Dorian Gray*, writes, "A cigarette is the perfect type of a perfect pleasure. It is exquisite, and it leaves one unsatisfied. What more can one want?" This would seem to be k d's view of love: an addiction that must inevitably disappoint.

It's an album she's been working up to for years. She's covered "I'm Down To My Last Cigarette" and "Three Cigarettes In An Ashtray" and, best of all, "Black Coffee", a smoldering Peggy Lee favorite that climaxes:

> *Now man was born to go a-lovin'*
> *But was woman born to weep and fret?*
> *To stay at home and tend her oven*
> *And drown her past regrets*
> *In coffee and cigarettes?*
>
> *I'm moanin' all the mornin'*
> *Moanin' all the night*
> *And in between*
> *It's nicotine…*

You get the idea. (That, by the way, is the first use of "nicotine" in a Tin Pan Alley lyric.) lang's best-known original song, "Constant Craving," speaks for itself. Her best-written song, "Miss Chatelaine" from the CD *Ingénue*, is about love as an exhilarating high, and the next track on that album begins, "You swim through my veins…" Artistically speaking, k d lang is addicted to addiction.

She's not the first, of course. Harry Warren and Al Dubin covered most of the bases in a song for *42nd Street* in 1933:

> *Ev'ry kiss, ev'ry hug*
> *Seems to act just like a drug*
> *You're getting to be a habit with me*

Let me stay in your arms
I'm addicted to your charms
You're getting to be a habit with me
I used to think your love was something that I
Could take or leave alone
But now I couldn't live without my supply...

Only in the final eight bars do the writers pull any punches:

I must have you ev'ry day
As regularly as coffee or tea...

Tea? You mean, after all that, we're talking about hot beverages here? Well, probably. Coffee's addictive; so are reefers (Cab Calloway, Fats Waller) and coke (Cole Porter – "I get no kick from cocaine"). But, in most pop songs, cigarettes are merely a stylish accessory.

The most stylish smoke comes at the opening of one of the most recorded songs ever. Eric Maschwitz, a BBC radio producer moonlighting under the name Holt Marvell, wrote the lyric as an attempt to come up with his own "You're The Top"-like laundry list. As it happens, I think he improved on the original. His is one of the most quoted lines in all popular music:

A cigarette that bears a lipstick's traces
An airline ticket to romantic places
And still my heart has wings
These Foolish Things
Remind me of you...

In the cold light of day, it's an image that's as likely to make you feel icky as nostalgic - a half-smoked fag with some cheesy lip gloss on it. But, set to those notes, it's a fine example of the transformational properties of music - the perfect opening for Maschwitz's rueful accumulation of sophisticated memory-joggers – "wild strawb'rries only seven francs a kilo," "the waiters whistling as the last bar closes," and so on.

Since then, the singing cigarette has dwindled away to isolated outposts of adolescent rock, like "Smokin' In The Boys' Room." One of the consequences, for an album like *Drag*, is that the most innocuous songs now, paradoxically, pack more of a punch than ostensibly more searing stuff like "My Old Addiction." k d lang is never more brazen than when dusting off "Smoke Rings," from 1932:

> *Puff! Puff! Puff!*
> *Puff your cares away*
> *Puff! Puff! Puff!*
> *Night and day...*

Lang's great quality is that she can pull off even the most anachronistic trifle without patronizing it. The songs emerge as charming and dated yet somehow contemporary. Her approach, which she's used consistently since recording Cole Porter's "So In Love" seven years ago, is to honor the broad parameters of the original layout - tempo, arrangement - while using pared-down, guitar-based rock orchestrations. The only surprise is that more singers haven't picked up on it.

I'd be interested to see if you could apply that technique (and the arrangements) to older material such as Victor Herbert's 1905 nod to Kipling, "(A Woman Is Only a Woman) But a Good Cigar Is a Smoke" - though that's probably an unlikely sentiment for k d. Still, it's these older songs that seem the correct assessment: Smoking is a consolation for the vicissitudes of life, a prop for losers. That's how Sinatra's been using cigarettes for 60 years, dimming the lights and getting them out for "Angel Eyes" or "One For My Baby," sad songs for guys with nothing to do but drown their sorrows. It's a persona that found its apotheosis in 1981 on the cover of *She Shot Me Down*, his darkest album: Sinatra, hunched in black leather jacket, wreathed in smoke, Jack Daniel's in hand. It's not an image the tobacco companies care for. When a British company, back in the Sixties, tried the saloon-singer approach in a TV ad - with a moody, reflective loner and the tag line "You're never alone with a Strand" - it wound up putting itself out of business. The Joe Camel approach is unlikely to produce any decent songs, but it sells more cigarettes.

Whatever happens to tobacco sales under the new settlement between the states and the tobacco companies - or, more accurately, between the tobacco lobbyists and some enterprising tort lawyers – it's bound to result in the further withdrawal of cigarettes from the mainstream of popular culture, including songs. That being the case, maybe, like the fatalistic protagonist of Victor Herbert and Glen MacDonough's 1908 operetta *Algeria*, we should light up one last time:

> *Fragrant clouds then from us veil*
> *Ev'ry sorrow, ev'ry doubt*
> *Till we wake at last to find*
> *That our cigarette is out.*

TEETH

Brush off

February 25th 2001
The Sunday Telegraph

ASKED WHETHER he and his British guest had discovered anything they had in common, George W Bush took the "special relationship" to a new intimacy. "We both use Colgate toothpaste," he replied. Say what you like about Tony Blair and Bill Clinton's much vaunted friendship but they had entirely separate views on oral relief.

Perhaps the inauguration of this new transatlantic bond was followed by both leaders unveiling the ceremonial plaque, but the press reports are unclear on this point. At any rate, it was in its own way an observation of ingenious ambiguity. For the viscerally anti-American Old Labour old-timers, the chaps who spent the Eighties demonstrating outside US bases in the United Kingdom, it is, of course, a perfect distillation of the essentially colonial nature of the special relationship: the British Prime Minister agrees to have Yank dental paste stationed on his teeth. For some, it's simply an acknowledgment of political reality: Tony's gleaming molars, like so many of the old country's military assets, may be nominally British choppers but, without American operational support, they'd be just another symbol of post-imperial decay.

For others, the image of Bush and Blair's toothbrushes side by side in the glass at Camp David sums up the faux intimacy of the special relationship: after all, according to Colgate-Palmolive, nearly four out of ten people around the world use Colgate toothpaste. At April's Summit of the Americas in Québec City, Mr Bush will be surrounded by fiery Latin *presidentes* neutralising the incendiary whiff of their enchiladas with the tangy freshness of Colgate. Indeed, when one takes into account impoverished jurisdictions where the peasantry's still squeezing the bottom of the tube they spent a week standing in line for in 1978, I would wager that Colgate usage in the world's presidential palaces is considerably higher than 40 per cent. And one reason it controls 40 per cent of the toothpaste market is because of the amazing range of its products: it even makes Colgate Star Wars toothpaste, mainly for impressionable youngsters but no doubt equally appealing to Republican Presidents hung up on missile defence. They also make Colgate 2-in-1, just the thing for a British Prime Minister who in his keynote address to the Canadian Parliament was eager to assure his hosts that the choice between traditional North American allies and newer Euro chums was a false one and that Britain could have both. Yes, the genius of Colgate is that it makes something for everyone, even Saddam (Colgate Loony Tunes toothpaste).

But toothpaste-wise Mr Blair's mission was one of Colgate Tartar Control. That's to say, the British feared that the new President would wave ta-ta to Mr Blair, writing him off as a Third Way has-been left over from the Clinton era, to be consigned to the trash can of history along with other icons of the age like Monica and Paula. This was a legitimate concern. Mr Blair got minimal media attention during his visit. If it's any consolation, so did Mr Bush. Instead, the front pages are all full of Mr Clinton, apparently still in the early stages of his long goodbye. You'd never know Dubya was President, which apparently suits him just fine. But, for a Prime Minister with more cosmetic concerns, it marked quite a difference: at least when he was Bill's armpiece, he used to get shown off around town and taken to parties.

Otherwise, as the President said, the Prime Minister "put the charm offensive on me. And it worked." From Mr Clinton, this would have been the usual schmooze. But Mr Bush genuinely likes pretty much everyone: slogging around New Hampshire last winter, he seemed to like me; but then again he's fond of Jesse Jackson, who's spent the last three months comparing him to Nazis and slave-owners. Dubya is a dating agency's dream client: he's open to everyone.

"The Prime Minister informed me this morning that he exercised at the gym prior to meeting Vice-President Cheney. I informed him, after this press conference, I'm going to go exercise in the gym," he announced. But just because one likes exercising in the morning and the other in the afternoon doesn't mean they haven't got a lot in common. "We both got great wives," said the President. Also, they both recognise that "our most important responsibility is to be loving dads".

Mr Blair grinned his Colgate grin. "That's enough to be going on with," he said.

I suspect Dubya genuinely likes Tony, and Tony genuinely thinks Dubya is a boob. It was, as Washington has learned in the last month, typical Bush.

The President, unlike his preening predecessor, is a personally modest man, happy to let the other fellow feel he's come out ahead. Two headlines nicely captured this approach: "Bush Supports European Rapid Reaction Force" but "Blair Stops Short Of Backing Bush Missile Shield Plans". In other words, both men have had to grit their shining teeth on each other's pet project but Blair didn't have to back down as much as Bush.

That's not how Washington views it. They don't care for the embryo EuroArmy, but they see it not as an American problem so much as a British one - as Mr Blair or his successor will eventually recognise. And Mr Bush, in his deceptively casual way, found a formulation that puts this ball firmly in the Blair court: "He assured me" that there would be a joint command; "he assured me" that the planning would take place within Nato. In Washington, they're well aware that these "assurances" are explicitly at odds not just with what the Continentals are saying but with the Nice Treaty itself.

Traditionally, this is the position Her Majesty's Government takes with its own subjects on European matters, "assuring" them that the black-and-white print – Maastricht's inauguration of "European citizenship", for example - is meaningless and to pay no attention to it. But Mr Blair's "assurances" are the first time the Americans have been subjected to bog-standard British Eurodissembling, and they're not fooled by it.

The Rapid Reaction Force, said the Prime Minister, is for situations where the Americans choose not to be involved. And just which situations did he have in mind? Ten years ago, the US wanted to sort out Yugoslavia, but the EU told them to butt out. "The hour of Europe has come!" declared Jacques Poos, the lion of Luxembourg. The hour of Europe came and went, and a quarter of a million corpses later the EU was only too glad for America to butt back in. An instinctive non-interventionist, Mr Bush sees no reason to intervene in the next designated Europetard on which the EU intends to hoist its ambitions.

Conversely, on missile defence the President is going to go do it anyway, but if it makes Mr Blair happier to pass off a *fait accompli* as a "debate" that's "important to have" and "take forward in a constructive way", it's no skin off Dubya's Colegate-shiny teeth. NMD is a much more important development than the Rapid Reaction Force, and the movement is all Bush's way.

This was made clear by the typically witless remarks of the buffoon who currently occupies the Canadian Prime Minister's office, who in the course of Mr Blair's visit insisted that he knew Bush's thinking and Washington wouldn't go ahead if the Russians and Chinese weren't happy. The guy was forced to retract, but his ineptness did usefully clarify the stakes: all Britain really has to do is make sure it's not the last Nato country to get with the programme, and given the behaviour of Canada, France, Germany, etc, there was never any real danger of that.

Instead, Mr Blair's gone further than he had to, and conceded that, if and when NMD comes, the Americans will be permitted to upgrade their UK radar installations, thus bringing Britain within the system. Obviously, if the Labour Party and the many opponents of NMD in the British press are determined that the United Kingdom remain vulnerable to nuclear annihilation, the Bush Administration would be willing to promise not to tell Downing Street when the Fylingdales radar station detects incoming missiles targeted on Birmingham. But, pending that memorandum of agreement, the missile shield is under development and Mr Blair understands that Britain will be covered by it.

The Prime Minister could have been more enthusiastic. He might have said that, as a "moderniser", he shared Dubya's view that a treaty negotiated by Brezhnev and Nixon is no more "relevant" to the new millennium than the House of Lords or a bunch of toffs in hunting pinks. But that would have been expecting too much of Tone's modernising zeal, wouldn't it? In the end, for all their fondness for Colgate, he's a pink toothbrush and Dubya's a blue toothbrush. One "moderniser" is obsessed with image and symbols and petty class grievances; the other is rethinking the fundamental assumptions of the last half-century.

From Colgate's point of view, the chipmunk-cheeked Prime Minister is probably a better testament to its fine products than the President, whose "smirk" was much criticised during the election campaign. But, from a geopolitical standpoint, I would say it's Mr Bush who has the Colgate ring of confidence about him. And, in his own way, the President managed to force a lot more than the Camp David toothpaste down the Prime Minister's throat.

MOUTH

**** *** ******!
,   .

March 2nd 2002
The Daily Telegraph

Gladiator's Four-Letter Threats To TV Director – The Daily Mail

Mottram's Oath-Laden Prediction – The Daily Telegraph

'VE WATCHED with envy all this week as my colleagues Robert Harris, Alexander Chancellor and Alice Thomson have peppered their columns with a ton of expletives deleted. Fortunately, just as I was wondering how I could match them, I got a call asking if I could host this year's Riskie Awards at the Grosvenor House. As you know, each year the Riskies honour the foul-mouthed men and women who've given us the sound bites with the most asterisks in them.

"Live from Park Lane," I began, "it's an asterisk-studded gala of four-letter favourites!" The Syd Lawrence Orchestra played Snoop Dogg's "Shut Tha F*** Up, Muthaf******" as my co-host, curling queen Rhona Martin, walked out with her corn-broom. "You know," she said, "I'm not really sure why I'm here tonight as the only swear words I ever use are 'You daft brush!'"

The audience sat on their hands and glared. There was a flurry of muttered asterisks from the front table, so I thought it wise to get down to business. "And the nominees for this year's 2002 Riskie for the most asterisks in a single sound bite are…"

"Star of stage, screen and banqueting suite ante-room, Russell Crowe for his post-Bafta small talk," said Rhona, introducing the video clip:

"I don't give a f*** who you are. Who on earth had the f****** audacity to take out the Best Actor's poem? You f****** piece of s***. I'll make sure you never work in Hollywood."

"Russell did stumble badly in his finale," I explained, "going for the technically difficult triple-f***, but losing his balance and limping through to the end without managing a single asterisk in his last sentence. However, his flawless performance in the early stages was enough to garner him his fourth nomination."

"Garner this, you f***!" said Russell, lunging at me and jabbing his finger in my eye as Rhona continued.

"Our next nominee is all-time champ Mike Tyson, back this year for a little light banter with a gentleman from the media at his press conference with Lennox Lewis."

"I'll put your mother in a straitjacket, you punk-a** white boy. I'll f*** you in the a** till you love me, f*****. You're a little white p**** scared of a real man. You wouldn't last two minutes in my world, b****."

"This was a flawless technical performance, though some judges did feel it was too similar to last year's routine," Rhona noted. "Our third nominee is Sir Richard Mottram, permanent secretary at the DTLR."

"It's the novelty hit of the year," I said. "You know the words, all together now:"

"We're all f******. I'm f ******. You're f******. The whole department's f******. It's been the biggest c***-up ever and we're all completely f******."

"And finally," said Rhona, "with her first ever nomination, Her Majesty The Queen for her speech in the New Zealand parliament."

"I am delighted to mark f**** years in the life of this country, and f**** years as Queen of New Zealand. Whenever P***** Philip and I are here, we particularly appreciate the f*********** of New Zealanders; the diverse c****** and the s******* b***** of the country; and the opportunity to share your s****** way of life."

"And now," said Rhona, "while Russell reads 'Trees' by Joyce Kilmer, we're going to drown him out with a medley of this year's nominations performed by Céline Dion and Tony Bennett!"

"You're f******," sang Céline.

"And you're f*****," sang Tony.

"The whole department's f*****," they sang together, Céline adding several appealing melismas – "f*-**-**-**-**-**-*". But I could hardly hear them over the rising voices backstage. I hurried off to see what was going on, only to find Russell, Mike and Sir Richard clustered round the Queen.

**** ***, ******!

"I'll make sure you never work in Hollywood!" yelled Russell, jabbing his finger in her tiara and getting a nasty cut.

"Oh, yeah?" scoffed Her Majesty. "Well, I'll make sure you never work at Sandringham!"

"What's the problem?" I asked.

"She's faking her asterisks," complained Tyson. "That bit about 'the f*********** of New Zealanders', it's some punk-a** p**** word like 'friendliness', not the real f-word."

"It's the biggest c***-up ever," fumed Sir Richard. "The whole show's completely f*****."

"Oh, put a s*** in it!" I snapped.

"There you are," said Russell. "Now you're doing it. Your missing word is 'sock', you pathetic piece of s***. 'Sock' isn't the real s-word."

"Who says?" I said. "You guys have been talking in asterisks so long you've no idea what you're saying."

"You mean, when I called that BBC bloke a f****** piece of s***, I meant he was…"

"A fleecey piece of sock," I said.

"Geez," said Russell, a little subdued. "No wonder he's not suing."

"We're all f…" began Sir Richard.

"Yes?" I said, encouragingly.

"We're all f… f… *feeble*. I'm *flirty*. You're *fleshy*," he trilled, playfully pinching Mike. "The whole department's *freaks* and we're all completely *flawed*! By George, I think I've got it!"

But by now the Queen was exchanging pleasantries with the wardrobe mistress. "Have you c*** f**?" she asked.

"Let me do that one," said Tyson, eagerly. "'Have you come far?'"

TONGUE

Sarnie

August 19th 2000
The National Post

I DON'T KNOW which focus group we have to thank for it, but with hindsight The Kiss was the masterstroke of that Democratic Convention - the "kiss heard 'round the world," as my colleague John O'Sullivan put it. I'm talking about Al kissing Tipper, I hasten to add, not Joe Lieberman kissing off his principles, a far more perfunctory osculatory manoeuvre: just a quick peck on the cheek and every single one of 'em - school vouchers, affirmative action - were bundled on to the Greyhound with a one-way ticket. By comparison, Al and Tipper's lip-lock was a work of art. The Vice-President entered the Staples Center, made his way through the crowd pressing the flesh, and then bounded up on stage to *really* press the flesh with the lush, sensual big Tipper. For what seemed like a good ten minutes, they "sucked face" (in the American vernacular) with a prolonged "tongue sarnie" (in the British). It told the American people everything his speech couldn't. Read his lips: No new interns.

Al, an obsessive micromanager when it comes to policy and politics, has decided to delegate when it comes to image: You want a beacon of morality? That's Lieberman's department. You want a regular guy? Check with my old roommate, Tommy Lee Jones. But there are still one or two things he doesn't need a body double for.

Tipper played her part, too, and not just in her drum solo during that Grateful Dead number. In that kiss, long and deep and (at least by comparison with the last eight years) very real, it was almost as if she was hoping through sheer passion to transform her frog into a prince. Al looked Grateful, but was he still Dead? The polls and pundits say no. He's a man reborn.

Going into the convention, Al had somehow contrived to have a gender gap with both genders. Dubya was way ahead with men and women, and that left Al running out of genders to have a gap with. Miraculously, after the convention, Al leapt into a commanding lead among women. The pundits attributed this to the speech - the stuff about HMOs, prescription drugs, Al's reaffirmation of "a woman's right to choose." The Vice-President is nothing if not enthusiastic on the last point – up to and including such hitherto arcane "choices" as the right of a pregnant woman on Death Row to take her foetus into the lethal injection chamber with her. No "safe, legal and rare" Clintonian ambivalence here. Al is positively gung-ho about its virtues.

Still, I think the experts are in danger of overestimating the bloodthirstiness of the average soccer mom. From my own unscientific researches, I'd say that what bumped Al up with female voters was not the speech but the kiss. Back in '88, the media liked to sneer that George Bush reminded every woman of her first husband. Al has something of the same effect, but that kiss was a sweet reminder that sometimes first husbands are worth sticking with.

The question, however, is how far should he go. In one of the heartwarming videos shown at the convention, Al drew the camera's attention to a nude self-portrait by Tipper. She was pregnant, thus testifying not just to her lush full-breasted loveliness, but to the potency of Al's seed. Now I don't doubt that Al and Tip are one hot couple. Statistically, the evidence favours this conclusion: According to a massive 1994 study, the Americans getting the most, the best and the liveliest sex are ...boring old monogamous spouses. The swingers cruising the bars are going to a lot more trouble for a lot less action. Anecdotal evidence also supports the stats: Al's sex life could hardly be worse than his boss', with his perfunctory un-"completed" encounters while he's on the phone to senior Democrats discussing Congressional business. But that doesn't mean that on the campaign trail the electorate wants Al and Tipper to go beyond a little light necking. Not so long ago I asked a Hollywood exec how he thought the supposedly erotic *Eyes Wide Shut* with Tom Cruise and his Aussie spouse Nicole Kidman would do. "It'll bomb," he said. "Audiences don't want to see married people getting it on."

It was while sitting under Tipper unzipped that Al wrote his eco-tome *Earth In The Balance*. Getting Gore balanced is the trickier proposition. He has a tendency to overcompensate. In the primaries against Bill Bradley, he was told he was too complacent, so he became too vicious. In the general election against George Dubya, he was told he was too cold, so he's become too hot. And, when his tongue's not busy probing the deepest recesses of Tipper's gullet, it tends to betray what really turns him on. He wrote his peroration himself, as most members of Local 573 of the Amalgamated Union of Speech Writers could tell. The more Gore tried to connect with "ordinary people," the more they sounded like a remote Amazonian tribe on a Discovery Channel documentary: For a quarter-century, he told us, "I have been listening to people, holding open meetings, in the places where they live and work. And you know what? I've learned a lot." No professional speech writer would have let that "they" slip by. "You" would have been better. "We" would have been best. "They" sounds like he's giving a debriefing back to his masters on Planet Zongo: "The places where they live and work are known as 'apartments' and 'donut shops.' After close observation, I have learned that in order to breathe they require a constant intake of 'triple-thick shakes'. For nutrition, they eat the flesh of an animal called the 'McNugget.' I will remain on Planet Earth awaiting further instructions."

So, after one week, how's Gore doing in his efforts to "connect" with earthlings? A couple of days ago, his Mississippi riverboat pulled up at a lock at Seneca, Wisconsin, and he bellowed at the crowd, "Tell me about your community!" There was an awkward silence.

Then someone mentioned the local championship girls' volleyball team. Al was suddenly pumped. "So y'all support Title IX?" he roared.

The townsfolk looked blank. Was Title IX perhaps the name of Al's volleyball team back in Tennessee? No, it's the federal legislation requiring equal funding of girls' sports. Go, Title IX, go! What a loss for sports fans that Al never applied for the Monday Night Football gig. The score so far: Title 9, Al zip.

For a "populist barnstormer," Al's recently installed software program could use a few modifications. But that's what happens when you learn to simulate human behaviour by memorizing the relevant federal statutes. To be honest, much as I adore Tipper, I'd rather see a nude portrait of Al: Is that his butt or the slot where you insert the CD-ROM?

My contact inside the Governor's mansion in Austin tells me that Dubya's a bit unnerved by Al's convention bounce and is wondering whether he should step up his gruelling schedule of campaign appearances to, oh, maybe three a week. Don't panic, Governor. The numbers of the "populist barnstormer" will settle down in a week or two. As those of us in Los Angeles were all too aware, many delegates traipsed glumly out of the Staples Center, privately conceding the convention had been a disaster and sounding a lot like Republicans in '96 - stuck with the party-establishment candidate because it's his turn, wishing the guy could be more like his wife, clinging to the hope that the other fellow will go belly up. The Tipper kiss was great, but, to adjust the priorities of the song, you must remember this: a kiss is just a kiss, a sigh is still a sigh. And, notwithstanding any upticks in the polls, as the Dems left Los Angeles that sound you heard was one long, weary, resigned sigh.

THE VOICE

Sinatra

May 17th 1998
The Sunday Telegraph

ON HIS 1979 futuristic concept album, *Trilogy: Reflections Of The Future In Three Tenses*, Frank Sinatra had it all figured out:

And when that cat with the scythe comes tugging at my sleeve
I'll be singing as I leave...

By the time the cat with the scythe finally showed up on Thursday night, Frank wasn't singing too much any more. But he still owed Capitol Records a new album - and that's all you need to know: he was the only 82-year old pop singer with a contract with a major record company. He did it his way, and he did it longer and better than anybody else in popular music. He was in the Hit Parade in the Thirties and in the Nineties. He started singing with Harry James and Tommy Dorsey, and, by the end, there was no one left but Pavarotti and Bono from U2, so he sang with them, too.

"Rock'n'roll people love Frank Sinatra," said Bono at the 1994 Grammy Awards, "because Frank Sinatra has got what we want. Swagger and attitude. He's big on attitude. Serious attitude. Bad attitude. Frank's the Chairman of the Bad." If only 20 per cent of the gossip is true, it was an amazing life: Frank delivering two million bucks in an attaché case for mob boss Lucky Luciano; the horse's head left in an uncooperative producer's bed; Nancy and Ava and Lana and Marilyn and Lauren and Mia in his bed, being very cooperative, sometimes (Ava and Lana) simultaneously; even towards the end, when ex-wife Mia Farrow told him of her troubles with Woody Allen, Frank sportingly offered to break Woody's legs.

But what's even more amazing than the life is that the records live up to it, and then some. The swagger and attitude, the chicks and mobsters are the incidental accompaniment; the real drama is in the songs. A couple of weeks ago, I was driving out of New York one Friday afternoon when "Come Fly With Me" came on the radio. Billy May wrote the arrangement while drunk, an hour before the recording session, and it still dazzles: the perky intro sort of taxis down the runway and the band lifts off, and there's Frank - and as you hit the ramp to the Bronx River Parkway, the skyscrapers fall behind you, the highway clears, and all the possibilities of America lie ahead.

When the song ended, I thought of the last time I'd spoken to Sinatra, a few years back. We were talking about Billy May, but Frank was fairly exhausted and distracted and rambling. Everyone in the room wanted a piece of him, and he was finding it hard to concentrate. But a couple of days later there he was on stage, pushing himself through punishing art-song arrangements of "Lonely Town", that long, complex Bernstein melody from *On The Town*, and the Soliloquy from *Carousel*, the role he should have played on film but instead played for ten minutes a night in recital halls and sports arenas around the world for half-a-century. Round about that last time we met, I saw some guy sing the Soliloquy in the National Theatre revival of the Rodgers and Hammerstein show: great voice - if you think a voice is about hitting notes and holding them for the requisite length. But the fellow had nothing to say. Sinatra, a couple of years shy of 80, could still make you believe he was a leathery old roustabout, scraping a living along the Maine coast, contemplating the birth of his first child.

Of all the pop idols, from Jolson to Madonna, who've crossed over to movies, Sinatra was easily the best. Why would you be surprised by his Oscar-winning performance in *From Here To Eternity* (1953)? This guy learnt everything he needed to know about building character and creating a role from Rodgers and Hart, Harold Arlen, Johnny Mercer. That's one reason why, despite a film career that's little more than a handful of high spots buried between kiddie musicals in the Forties and some shrugged-off Rat Pack capers in the Sixties, Frank's had more influence on American film acting than most full-time thespians. In the Fifties, he became the first dark celebrity, the glamorous outsider, mostly fatalistic, occasionally bitter, and potentially easy to provoke. He shared more than his hat brims with film noir. In fact, one of the best examples of film noir is his 1981 album *She Shot Me Down*, an accumulation of regrets, reflected through neon bar signs and drizzled sidewalks and dying streetlights.

He poured so much of himself into the music that it drained everything else. By the end, he'd given up the high-rolling, the really late nights, the broads and the booze and the cigarettes; he could barely talk. Even "It's a privilege to be back in your wonderful city" seemed an effort. But, week after week, on stage in some grim rock stadium on the edge of a shopping mall in some nondescript suburb, the lights would dim and Sinatra would effortlessly shrink the place to the size of those poky, smoky New Jersey saloons of his youth. Just a voice and an old saloon song in salute of some emblematic long-lost loser whose "chick split, cleaned out his stash and left him with a room full o' nothin' cryin' into a gallon o' Muscatel" - and underneath him Bill Miller, his long-time accompanist, would begin the bar-room piano intro to "One For My Baby (And One More For My Road)". The rock crowd – "the boys in the funky shoes", as he called them - needed dry ice and laser shows to fill the place; Frank made do with the old props - a tumbler and a cigarette.

It was always about the music, even in the beginning. Before Sinatra, male singers aspired to the condition of Bing Crosby, who sang like he played golf: let's knock it around for a while and get to the clubhouse without breaking into a sweat - 18 holes, 32 bars, hey, what's the diff? When Crosby sang, that was all he did: sing. If you want an extreme example, try "Home On The Range". Crosby's recording says nothing other than, "Ah, let's gather round the old joanna and sing a backporch favourite from 1873.' Sinatra's version is extraordinary: his home really is on the range, and the deer and antelope are frolicking about 15 yards from the microphone.

Inevitably, there soon was heard a discouraging word: Frank Sinatra sings songs, said one early reviewer, as if he believes them. And he meant it as a criticism. Maybe Stanislavskian identification isn't such an advantage when you're doing a song about wanting a home where the buffalo roam, but, as Sinatra emerged from Crosby's shadow, it was to prove decisive - and, in the long run, render Bing's approach inadequate.

The Thirties crooners bounced thoughtlessly through romance and glossed over pain. For Sinatra, that was not enough. "Take 'Fools Rush In'," he said, by way of example. "The story of the song is in the first four lines: 'Fools rush in where wise men never go, but wise men never fall in love, so how are they to know?' But most fellers chop it up: 'Fools rush in (breath) where wise men never go (breath), but wise men never fall in love (breath), so how are they to know?' But, if you do it that way, nobody follows the story." He learned breath control from swimming underwater at Demarest High School in Hoboken and later from Tommy Dorsey's trombone. Sinatra used to sit on the bandstand behind Dorsey, wondering why the bandleader's jacket never moved, as it surely would if he was taking even the slightest breath.

But for someone who represents the apogee of popular singing, he was never really, apart from that first flush of shrieking bobby-soxers back in the Forties, a pop singer. Pop is fashion, and Sinatra was usually at odds with whatever the prevailing fashion was. When pop singers were regular guys like Bing, Frank was spilling his guts out and introducing to the Hit Parade such fine emotional niceties as self-disgust. When Eisenhower's America promoted cosy, domestic, picket-fence family values, he recast himself as a ring-a-ding, swingin' bachelor. At 50, a time when most celebrities are still pretending they're 28, Sinatra embraced premature old age and songs of wistful regret: "(When I Was 17) It Was A Very Good Year".

Jerome Kern once gave the young British composer Vivian Ellis a piece of advice: "Carry on being uncommercial. There's a lot of money in it." It worked for Sinatra. Back in the Fifties, the smart money was on Mitch Miller, head honcho at Columbia Records, the man who single-handedly produced the worst records of the era and so debauched the currency of mainstream Tin Pan Alley that rock'n'roll seemed a welcome alternative. It was Miller who insisted that Frank record "Mama Will Bark", a doggy duet with the big-breasted Swede Dagmar, which, as the lyric puts it, is "the doggone-dest thing you ever heard". Sinatra quit Columbia but never forgave Miller. Years later, they chanced to be crossing a Vegas lobby from opposite ends. Miller extended his hand in friendship; Sinatra snarled, "Fuck you! Keep walking."

Fuck you! Keep walking: it could be the tempo marking on any one of those surging big-town swing arrangements. Indeed, his entire oeuvre could be sub-titled "Fuck you! Keep walking". Pop music never really deserved Sinatra, a man whose instincts have invariably been better than anybody else's. His first starring role was in the film version of Rodgers and Hart's *Higher And Higher* (1944). Then he discovered that the studio had slung out most of the score, including one of the finest ballads of the century, "It Never Entered My Mind". So he went and recorded it as a pointed rebuke to the studio mush-heads. It was to be one of the first "standards" saved by Sinatra. The very notion of a standard - a song that endures and can be re-investigated over and over across the decades - is a Sinatra invention. At the time, Mitch Miller would rather he'd sung "She Wears Red Feathers (And a Hula-Hula Skirt)".

In the Fifties, Sinatra moved on to invent the album, approaching it like a song-cycle, a dramatic journey. His new label, Capitol, would have been just as happy if, like Ella Fitzgerald, he'd simply plucked Irving Berlin's 20 biggest hits and called it a "Songbook". When you eavesdrop on some of his rehearsal tapes, you appreciate how much he contributes to his arrangements, tinkering with them again and again until he's satisfied. Recording "The Road

To Mandalay", he turned Rudyard Kipling into a finger-snappy swinger anxious to be back east of Suez where "a cat can raise a thirst". There was a 32-inch gong in the arrangement, which the percussionist Frank Flynn walloped on the line "And the dawn comes up like thunder..." after which Sinatra wrapped up the chorus – "out of China, cross the bay". But he had problems with the ending, so he told Flynn: "Next time round, really hit that mother." Flynn beat the gong, Sinatra picked up his hat, threw his coat over his shoulder and left the studio. The band fell around laughing and it took them about ten minutes to realise Frank had actually gone home. He knew that anything after the gong would be an anti-climax. The Kipling estate, in their infinite wisdom, had the record banned throughout the Commonwealth.

On the other hand, I've just been listening to "It's Sunday", a Jule Styne/Susan Birkenhead song from the early Eighties, with a spare guitar accompaniment. He gave it to several orchestrators - to Don Costa, who did Sinatra's definitive recording of "Come Rain Or Come Shine"; to Peter Matz, who does most of Streisand's stuff - but he was never satisfied. "Listen, I think you guys are missing the point," he said. "It's an intimate kind of thing." So he and his guitarist, Tony Mottola, went into the studio and showed them how it should be done. It's the only Sinatra recording with solo guitar, and it's beautiful: not a song for swingin' lovers, but a song for mellow grown-ups, for breakfast in bed with the Sunday papers.

I think it's great, but it wasn't great enough for Sinatra, and he wouldn't release it. He sings standards, and he has standards. In that sense, whatever its merits as a song, "My Way" is no idle boast. In the Seventies, the song became Elvis's first posthumous hit. What a joke: Elvis never did anything his way, only his minders' way. The record itself negates the title: he's stumbling through a song he barely knows in a generic Vegas schlock arrangement that sounds like a karaoke backing track for open-mike night at a Tennessee sports bar.

In the end, Mitch Miller gave "She Wears Red Feathers (And a Hula-Hula Skirt)" to Guy Mitchell and made him a star. But where's

Mitchell now? The vocalists touted as rivals half-a-century ago - Dick Haymes, Ray Eberle - died in obscurity. Only Sinatra endures, the consummate popular singer who survived rock'n'roll and lived long enough to see its leatheriest old rebels defer to his legend: in recent years, Bruce Springsteen would drop in on Frank to sing ballads round the piano; Bob Dylan kept pestering him to make an album of Hank Williams country songs. Stevie Wonder, Gloria Estefan and Luther Vandross joined him for a couple of gimmicky *Duets* albums. And, although those sets have their embarrassing moments, none of them is Frank's. The problem is his melisma-crazed partners, such as Patti LaBelle, elongating "lu-u-u-u-u-u-urve" to a seven-syllable word, as if soulfulness is measured by the yard. In this, as in other respects, the rock era has returned us to operetta posturing. Where operatic singing is nothing but generalised vowel sounds, in classic pop all the action is in the consonants. At 80, Sinatra was still the king of the consonants: nobody else got such a kickkkkkk out of "I Get a Kick Out Of You", riding the rhythm section in 4/4, the all-American time signature, what Nelson Riddle called "the tempo of the heartbeat". Frank was never a melisma man; as Jule Styne, his longtime composer and one-time flatmate, said to me: "Frank's figured it out. He sings the words. The other fellers sing the notes."

Sinatra was adored by his *Duets* partners for that swagger and attitude - the way he wears his hat, to quote Ira Gershwin - but, alas for popular music, he had no one to pass the hat on to. Towards the end, the gossip columnists wrote that his mental faculties were diminished. Wrong. It's the music business whose mental facilities are kaput, drowning Sinatra's beauty - Dylan describes his ballad voice as a 'cello - in the half-baked hoodlum exhibitionism of gangsta rap and the empty bombast of Céline Dion movie themes. Frank would have understood. If you want the entire history of pop music on one single - the tug between its highest aspirations and its basest instincts - Sinatra wrapped it up in 1951 on Columbia 39425: on the A-side, "Mama Will Bark"; and, consigned to the B-side, "I'm a Fool To Want You", an almost painfully raw rueful reflection which today ranks as one of his finest recordings. It is, supposedly, Frank spilling his guts out about his doomed love for Ava Gardner, but it survives long after Ava and that incendiary romance are forgotten.

So, if we have to make these songs a soundtrack, let's broaden it a little: it's not his story, but the century's. Compared with almost any rock act of the past 30 years, he could not have been less interested in using his talent for social comment. Yet somehow he's present at all the great events of our time. In the fall of 1989, Gennady Gerasimov, that slick western media-friendly spokesman for President Gorbachev, announced the new-look, ring-a-ding-ding Warsaw Pact: "The Brezhnev Doctrine is dead," he declared. "We now have the Sinatra doctrine: you do it your way." Back at the White House, Vice-President Quayle was encouraged by Mr Gerasimov's statement but, noting the continued presence of soviet troops in Eastern Europe, urged him to remember the Nancy Sinatra Doctrine: "These Boots Are Made For Walking". This is how Communism crumbled: not with a bang, or a whimper, but with Sinatra one-liners.

"Frank walks like America," said Bono. "Cocksure." Think of the opening titles of the film *Wall Street*: the commuter trains and ferries and buses and subways feed the workers into the city, thousands of them, swarming up from their subterranean tunnels and on to the sidewalks, anonymous stick figures dwarfed by skyscrapers. And above it all Sinatra sings:

> *Fly me to the moon*
> *Let me play among the stars...*

Without the song, the scene is nothing: for what is drearier and more humdrum than commuting? But Frank Sinatra walks not just for the pedestrian dreamers of Wall Street, but for the highest fliers. When Americans really did fly to the moon in 1969, the astronauts took Sinatra on their portable tape recorder and "Fly Me To The Moon" became the first song to be heard on the moon itself. Any other nation would have chosen the "Ode To Joy" or *Also Sprach Zarathustra*, something grand and formal. But Buzz Aldrin knew what the sound of our century is: what is the American dream but the breezy confidence of Frank Sinatra in 4/4?

> *Fly Me To The Moon*
> *Let me swing forevermore...*

You got it, Frank.

BEARD

No more close shaves

August 9th 2001
The National Post

I WOULD like to recant my endorsement of George W Bush. I got Campaign 2000 all wrong. For some reason, I spent a year and a half savaging Al Gore, whereas I now realize he is perhaps the most courageous politician in the Western world.

What has brought about my epiphany? Simple. Al has returned from his European vacation with an attractive beard.

No, no, I don't mean he's dating Nicole Kidman. I'm talking about facial hair. Al has come out of the closet and proclaimed solidarity with the most politically marginalized group in society: the bearded. The man I mocked as a weirdsmobile has become a beardsmobile; Mister Squaresville is Mister Hairsville. After last November, Al has decided he doesn't need any more close shaves. Good for you, Al. You grow, girl!

I should say at the outset that I have no personal interest in this subject. That barely discernible trace of five o'clock shadow in the picture that accompanies this column is an unfortunate side-effect from my time as East German ladies' shot putt champion. So I write simply as a neutral observer. But look at the G7 summit: from Bush to Berlusconi, not a whisker in sight at the top table. In the modern world, God forbid you should be racist, sexist, anti-Semitic, homophobic, but pogonophobia? We're up to our depilated ears in it. It's the last bigotry acceptable in polite society. Mrs Thatcher had huge numbers of Jews in her Cabinet, she kept the Earl of Avon in his ministry even when he was in the last throes of Aids, but it was an open secret in London that she would countenance no beards in government. If you're hairy and you want to be a Cabinet minister, book a one-way ticket to Kabul, where the Taliban have made beards compulsory, though whether just for men I'm not sure.

By contrast, the recent British Conservative Party leadership race was fought between a fatso, a baldie, a toff, a bi-guy and a dullard, but, as Fleet Street veteran Stephen Glover casually remarked, it's now impossible for a beardy to get anywhere at Westminster. The exception is the Home Secretary, David Blunkett, who's bearded but is also blind, and evidently the PC types feel a bit squeamish about demanding a guy who can't see scrape himself with razor blades every morning.

But, if you're not blind, what's your excuse? In 1992, Ross Perot ran for President as a man America could trust to do the right thing – that's to say, if elected, he would not hire homosexuals, adulterers or men with beards. This was undoubtedly a prejudice too far, since, at a stroke, he blew the inbred mountain-man vote, a natural Perot constituency but one that didn't take kindly to being consigned to the same roasting spit in hell as the fornicators and sodomites.

But, more importantly, although many commentators spoke out against Perot's blatant anti-homosexual stance, no one seemed the least perturbed by his outrageous anti-beardism. The only man to stand up to his insane campaign to tear out every last facial hair was an employee who'd converted from Catholicism to Orthodox Judaism and found himself sacked for the ensuing growth.

In court, Perot managed to find a rabbi to testify that, while the Bible prohibited shaving, it did permit plucking. Back on the campaign trail, the great man, in a clumsy attempt at hirsute outreach, pledged that, under a Perot regime, the congressional barbershop would be closed down. Presumably those guys could go pluck themselves, too.

Few leading figures are as ferociously anti-beard as Perot, but then they don't have to be, any more than previous generations had to be explicitly anti-Semitic or homophobic. The clean-shaven instinctively take care of their own: you shave my back, I'll shave yours.

Yet, once upon a time, beards were seen as a sign of authority, a necessary qualification for political leadership. In ancient Egypt, anxious not to look a wimp, Queen Hatshepsut went around in a false beard. The next couple of millennia were years of worldwide beard dominance, save for the rule of Peter the Great in Russia. The Tsar, who seems to have had some deep-rooted masculinity issues, loathed beards, and went around pulling out his servants' facial hair or shaving them so roughly that their skin came off. By 1698, he'd had enough. On September 5th, he personally shaved every senior courtier. "At a stroke," wrote his biographer, "the political, military and social leaders of Russia were bodily transformed. Faces known and recognized for a lifetime suddenly vanished." Peter then introduced a tax on all Russians who wished to remain bearded. They could do so, but only if they wore bronze medallions around their necks bearing a picture of a beard and the inscription "Tax Paid". At present, beards are untaxed in the US, but a small duty on selected parts of Appalachia could easily fund a prescription drugs program for the whole of Florida.

But Peter the Great's reign was an isolated incident. Where did the massive anti-beardism of contemporary society come from? Like so many other malign trends of the modern world, is it perhaps Canadian? America has Abe Lincoln and Ulysses S Grant and good old Uncle Sam, but Canada's icons are the most clean-cut in the world: from Wolfe and Montcalm to Macdonald and Cartier to Clark and Duceppe, where's the hair? Canadian politics was depilated long before any other western nation: while Britain flourished under the magnificent shrubbery of the Marquess of Salisbury, we had the arid plains of Wilfrid Laurier. Perhaps this is another confirmation of the thesis explored the other day by Michael Valpy in *The Globe And Mail* – that Americans are the masculine in North America, Canadians the feminine.

Yet, even in America, no politician has been elected to the White House with the old face fungus since Benjamin Harrison in 1889. That's what makes Al Gore's move so heartening to those of us ruthlessly excluded from the corridors of power by systemic discrimination. To be sure, his is a tentative beard, small, trim, the sort that could easily be mistaken from a distance for a faint smudge from an oversized Ben & Jerry's caramel cone. In gay terms, it's a first step, a shy glance that says "Bi-Curious Seeks Similar". But in years to come it will be recognized as a landmark in beard liberation.

To my surprise, I find that I'm the only columnist on the planet who recommended Al Gore grow a beard. It was a year ago, in America's *National Review*, that I advised Al, after he lost, to "grow a beard, grab that double-bladed axe and your best mule, and head for Montana". (I had noted that his views on the environment are indistinguishable from the Unabomber's, in whose cabin was found a heavily annotated copy of Al's eco-tome *Earth In The Balance*.)

And now, amazingly, he has taken my advice and grown a beard - not the full Unabomber, but on its way. Go for it, Al. Say it loud, you're hairy and proud. We're ravin', we ain't shavin', get used to it.

CHIN

Taking it on

October 10th 1992
The Times

IN THE SIXTIES, Joan Rivers and Norman Wisdom both appeared on "The Ed Sullivan Show". Unknown in America, Wisdom was turned down flat by Sullivan's agent, so, just like in the best backstage fables, he bluffed his way into the star's apartment and treated him to an unscheduled audition. This time, he got turned down by Sullivan himself and only cracked the show years later. Miss Rivers had an easier ride. Sullivan had booked the singer Johnny Rivers but, in a typically mangled sign-off, he announced: "Next week, we'll be having Joanie Rivers." So they felt they had to.

Each anecdote typifies its author, right down to Wisdom's misspelling of Sullivan's home as the Delmonca Hotel a charmingly parochial error which somehow emphasises the gulf between British show business and the real thing. In her memoir *Still Talking*, Miss Rivers talks a lot, mostly about struggling. But she falls into fame painlessly, and only later does it appear - to the reader if not to the author - that she's paid a truly hellish price for it. The cheery chin-up chappie of terrible post-war British comedies, Wisdom really struggled - abandoned as a child, uneducated, sleeping on the streets, running away to sea - but has been rewarded with a suburban contentment.

By contrast, in Hollywood, Miss Rivers seems to have wound up on the outside looking in. Most showbiz reminiscences are determinedly bland – in Norman Wisdom's *Don't Laugh At Me*, aside from a gripe about Tommy Cooper's smelly feet, the hindsight of Wisdom is generally benign. Bob Hope and Bing Crosby? "A prize pair of gagsters!" Miss Rivers, on the other hand, seems to have burnt whatever boats she had left long ago. Hope is the high-handed guy who dislikes funny women and "could not even say my name"; as for Crosby, "he was a drunk who screwed around and beat up his wife". And they're just passing shots.

Much of her book is devoted to settling scores with various Fox executives responsible for cancelling her talk-show. What's left is an unintentional guide to the exquisitely refined etiquette of modern celebrity. I especially enjoyed the painstaking reconstruction of the preparations for a benefit for battered children co-hosted by Miss Rivers and Elizabeth Taylor, when each star had the other's house staked out by security guards with walkie-talkies so that her limo would not be the first to arrive and have to hang around for the other's big entrance. Miss Rivers is famous for her lardbutt Liz jokes, but her victim sportingly takes it on the chins ("Liz Taylor has more chins than a Chinese phone book").

Joan's chin seems to have got lifted up and out of sight during her many and extensive cosmetic surgeries, along with a large chunk of whatever humanity she started out with. After the suicide of Miss Rivers' husband, Miss Taylor relays a message through Roddy McDowell: "If Joan wants to call me, she can." But the first rule of show business is "Do call us, we won't call you", and Joan would not touch that dial:

Call her? Who the hell does she think she is? Gee, if I'm real polite maybe she'll even say something kind to me. That anger was the first emotion to penetrate my numbness.

Miss Rivers' account of her husband's death is exhaustive and confessional, yet strangely unmoving. For years, he was a staple of her stand-up routine. But between "Edgar" the stage prop and Edgar the real Mr Joan Rivers, there was a huge gulf. Refuting W C Fields' chosen epitaph that "I'd rather be in Philadelphia", Edgar Rosenberg, faced with the choice in a Philly hotel room, jumped into a bottle of pills. Not long afterwards, I saw Miss Rivers on stage in Neil Simon's *Broadway Bound* and, like the blue-rinsed matinee matrons who had flocked to see her, I assumed that her personal ordeal gave a special conviction to her portrayal of a middle-aged wife and mother ground down by a harsh life. But maybe she's just a swell actress.

The over-analysis of what she went through is standard daytime talk-show stuff, from "constructive denial" and "the value of a non-judgmental person". But she does take her support group on a cruise to the Greek islands. Trauma, denial, rehab. It's the modern celebrity's equivalent of the Orpheum circuit. In the process, the event itself gets as lost as her husband. Who knows what she really felt? Fury, at least in part. But, as told here, it's not about the suicide, it's about coping with the suicide – and in the end all it's really about is changing the act: "My husband left in his will that I should cremate him and then scatter his ashes in Neiman Marcus. That way he knew he would see me five times a week." What's revealing about the joke, though, is that it's not really about him or his suicide, but her shopping habits. In death, as in life, he's still only the feed.

Some years earlier, stricken by heart failure, Rosenberg had been rushed to UCLA Medical Center and whisked past the red tape. "That is what celebrity is all about," Joan declares in apparent seriousness. The combination of self-absorption and trivia are like a black parody of contemporary Hollywood. Even the amusing anecdotes are delegated: at a Vegas strip joint, "Suzie Midnight made my hairdresser eat a marshmallow stuck on her nipple."

Norman Wisdom, you feel, would be keener on the marshmallow than the nipple. There's not much sex in his book, although "a poofy sailor" called Dmitri does lunge at him late one night. Wisdom's (or his collaborator's) style is quaintly outdated and unconcerned about political correctness. Not many comedians would boast about "rubbing shoulders with Prime Minister Ian Smith in Rhodesia", still less of taking along a copy of Mr Smith's wretched book to be autographed. But don't laugh at him 'cause he's a fool. He's happily settled on the Isle of Man, where they know how to treat "poofy sailors", and where he always tucks in to Sunday lunch at one o'clock sharp: roast lamb, roast potatoes, carrots, mushy peas, rhubarb and custard. "That's what I call real food."

On the other side of the world, Joan Rivers sits down to a terrine of monkfish with a zucchini sabayon, protected by electronic gates and private guards, with an address book full of contacts but only one couple she says she'd ever eat with in her eat-in kitchen, agonising over whether her daughter Melissa has "suffered the burden of a celebrity mother". A bit late for that.

JAWS

of defeat

September 7th 2003
The Chicago Sun-Times

HERE'S A NEWS item from the too-lame-to-be-true category. But it is. It seems Bill Clinton and Mikhail Gorbachev have collaborated on a re-telling of Prokofiev's beloved children's classic, *Peter And The Wolf*. In the original, you'll recall, Peter and his friend the duck are out frolicking in the meadow when the slavering wolf shows up and embarks on his reign of terror. He gulps down the duck as his hors d'oeuvre, and has the cat lined up to follow. But fortunately Peter gets hold of a rope and uses it as a noose with which to muzzle the wolf and take him into captivity.

In the Clinton version, you won't be surprised to hear, Peter realizes the error of his lupophobia and releases the wolf back into the wild. The wolf howls a friendly goodbye. Which is jolly sporting of him when you consider that it's all our fault in the first place. "Forgetting his triumph, Peter thought instead of fallen trees, parched meadows, choked streams, and of each and every wolf struggling for survival," narrates our Bill, addressing the root causes and feeling the wolf's pain. "The time has come to leave wolves in peace."

No word on the fate of the duck. Is she left in peace? Or in pieces?

And so the 42nd President brings us full circle, back to where we came in, two years ago. On the Eastern Seaboard, the weeks leading up to September 11th 2001 were the summer of shark attacks. Jessie Arbogast, an eight-year old lad from Pensacola, Florida, had his arm ripped off, but his quick-witted uncle wrestled the predator back to shore, killed him, and retrieved the chewed-up limb from his jaws. In a thoughtful editorial, The New York Times came down on the side of the shark: "Many people now understand that an incident like the Arbogast attack is not the result of malevolence or a taste for human blood on the shark's part," explained the Times. "What it should really do is remind us yet again how much we have to learn about them and their waters."

In other words, we need to work harder to understand "why they hate us". Just blundering into their waters in ever more culturally insensitive bathing suits will only provoke the vast majority of non-violent members of the shark community to hate us even more.

Two years after "the day America changed forever", the culture is in thrall to the same dopey self-delusion it held on September 10th 2001: there are no enemies, just friends we haven't yet apologized to. The terrorist won't be a problem if, like young Jessie with the shark, we just give him a helping hand. Or, as the novelist Alice Walker proposed for Osama bin Laden, "I firmly believe the only punishment that works is love."

That's why America's TV networks have decided to sit out this week's anniversary. On the day itself, it was all too chaotic and unprecedented for the news guys to impose any one of their limited range of templates. For the first anniversary, they were back on top of things and opted to Princess Dianafy the occasion, to make it a day of ersatz grief-mongering, with plenty of tinkly piano on the soundtrack and soft-focus features about "healing circles". That didn't go down too well, so this year they've figured it's easiest just to ignore it. The alternative would be to treat 9/11 as what it was – an act of war – and they don't have the stomach for that. War pre-supposes enemies, and enemies means people you have to kill, or at least stop, or at the very least be ever so teensy-weensily judgmental about. And, in an age when Presidents rewrite *Peter And The Wolf* to end with Peter apologizing to the wolf, why should the network sob-sisters be any tougher?

Back in the Nineties Bill Clinton didn't exactly apologize to the wolf, but he turned a blind eye as the poor misunderstood fellow pounced on more and more denizens of the barnyard, whether at

American quarters in Saudi Arabia, or American embassies in Africa, or even American buildings in Manhattan, when the World Trade Center was hit first time round. In Prokofiev's *Peter And The Wolf*, the children learn to identify each musical instrument as a different animal. But no matter how loud Osama bin Laden blew his horn he couldn't get the Administration's attention. According to Richard Miniter's new book, after 17 sailors were killed on the USS Cole, the Defense Secretary Bill Cohen said the attack "was not sufficiently provocative" to warrant a response. You'll have to do better than that, Osama!

So he did.

President Clinton's new CD usefully clarifies his party's problem this election season: a significant chunk of the American people think the Democratic candidates feel the same way about the war on terror as Bill Clinton does about Peter's wolf and The New York Times does about Jessie's shark. And they reckon they know how that usually winds up. A couple of years back, a cougar killed a dog near the home of Frances Frost in Canmore, Alberta. Miss Frost, an "environmentalist dancer" with impeccable pro-cougar credentials, objected strenuously to suggestions that the predator be tracked and put down. A month later, she was killed in broad daylight by a cougar who'd been methodically stalking her.

"I can't believe it happened," wailed a fellow environmentalist. But why not? Cougars prey on species they're not afraid of. So, if they've no reason to be afraid of man, they might as well eat him. He's a lot easier to catch than a deer or elk.

You can object that America's enemies in this war are not animals, though the suicide bomber seems to me not fully human, either. But nor are wild animals merely the creatures of their appetites. They're also astute calculators of risk. Aside from the boom in Islamic terrorism, the 1990s was also the worst decade ever for shark, bear, alligator and cougar attacks in North America. One can note that there are more of these creatures than ever before - the bear and cougar populations have exploded across the continent. But there's also the possibility that these animals have not just multiplied but evolved:

they've lost their fear of man. Not so long ago, your average bear knew that if he happened upon a two-legged type, the chap would pull a rifle on him and he'd be spending eternity as a fireside rug. But these days it's just as likely that any human being he comes across is some pantywaist Bambi Boomer enviro-sentimentalist trying to get in touch with his inner self. And, if the guy wants to get in touch with his inner self so badly, why not just rip it out of his chest for him?

North American wildlife seems to have figured that out. Why be surprised that the wilder life in the toxic Saudi-funded madrassahs did as well? Each provocation, "insufficient" to rouse Bill Cohen, confirmed Osama's conviction that America was too soft and decadent for the fight. Two years on, the defeatist elites of our culture are still desperate to prove him right.

THROAT

Full throttle

December 1996
The American Spectator

HENNY YOUNGMAN puts it best:

Take my wife. Please.

There it is in four words, one of the most indestructible artistic traditions: the need of the creative soul - whether novelist or stand-up comic - to co-opt his spouse into the act and to saw her in two or more pieces. With Youngman, at least there's a punchline. By contrast, with Philip Roth, in his novel *Deception*, the line is: "I wouldn't say my wife's a tedious, whining, middle-aged drag ...but she is."

In *Deception*, Philip Roth creates a writer called "Philip" who has a studio in London and is married to a dreary actress called "Claire." At the time, Roth had a studio in London and was married to an actress called Claire. So much for the novelist's powers of imagination - which, in this case, seem largely confined to the scenes in which various exotic beauties from Eastern Europe descend on said studio and engage in increasingly convoluted bouts of sex with him. As Roth is virtually unknown in London, never mind Eastern Europe, I think it safe to attribute these passages to authorial invention or, anyway, wishful thinking.

Is he much thought about here these days? For my own part, I'm always happy, if I'm in a motel room at two in the morning and can't sleep, and I've missed the rerun of "Leave It To Beaver" dubbed into Spanish, to stumble across another showing of Ali McGraw and Richard Benjamin in *Goodbye, Columbus*. But, according to my own unscientific survey of American youth, while many have heard of the phrase "Portnoy's complaint," they've no very clear idea what it involves or who Portnoy is.

Claire Bloom could put them straight. Yes, that Claire - the dreary, middle-aged actress married to the writer named Philip. At a stroke, Miss Bloom's new book, *Leaving A Doll's House*, has revived Roth's decayed notoriety in what *People* calls "the most buzzed-about celebrity memoir of the season."

Claire Bloom isn't really a celebrity. You may remember her from Chaplin's *Limelight* or as Lady Marchmain in *Brideshead Revisited* or even from a stint on "As The World Turns", but her various achievements have never coalesced into an integrated public personality. She is one of the most beautiful women of our time, with exquisite bone structure and big, deep eyes. But it's an austere, distant beauty. "I have always been very private," she assures every interviewer, and yet she's come up with the most public evisceration of a famous author since ...well, most profile writers cite Nora Ephron's *Heartburn*, but that's an unsatisfying comparison: For one thing, Carl Bernstein was too busy with Nixon to get around to trashing Nora in print.

By contrast, the Bloom/Roth marriage has now been written up from both sides, and what's striking is how much more persuasive the solid working actress's showbiz memoir is than the great novelist's work of fiction. By comparison with his banal *roman à* Claire, her account has the sharper crises - as when Roth demands Bloom's teenage daughter be banished permanently from the house and dispatched to a youth hostel - and the more telling details: in the divorce settlement, he graciously returns her china, cosmetics, fax machine and all the various mementoes of their love, including the plastic figure from atop their wedding cake.

Female revenge is big at the moment. It's powered this fall's biggest movie - *The First Wives Club* - as well as last winter's - *Waiting To Exhale*. In both cases, theaters have been packed with women cheering on their sisters. But, as revenge dramas go, they're pretty tame stuff - and, in most cases, just strategies for getting their men back, or getting other men who are mostly the same as their predecessors. Even the scene where Angela Bassett dumps her husband's clothes in the street and torches them lacks the precision of the baronet's wife in England a year or two back: she took the best bottles from her philandering husband's wine cellar and left them as presents on the doorsteps of the neighboring villagers; then she went through his wardrobe and carefully cut out the crotch from every pair of trousers in every one of his expensive Savile Row suits.

After that, it surprises me that the so-called "women's audience" goes along with such lame affirmations of sisterly solidarity as *The First Wives Club* - or, come to that, that their leading ladies do. Over the years, the powder room at Actors' Equity has surely crackled with more robust tales. Imagine a *First Wives Club* with, say, Claire Bloom and Mia Farrow. Whatever you think of these women, they've changed irrevocably the public perception of their ex-husbands - or, in Mia's case, ex-co-adoptive-caregiver-living-across-Central-Park. For example, I've always been fond of Woody's *hommage* to his Bogart obsession, *Play It Again, Sam*, but, in light of his recent travails, there are now all kinds of unnerving associations in lines like "Here's looking at you, kid."

Though Miss Bloom was an unintended beneficiary of the split with Mia (in *Mighty Aphrodite*, she landed the role of Woody's mother-in-law which, in his recent work, had tended to go to Maureen O'Sullivan, Mia's real-life mom), she seems remarkably similar to Allen's characterization of Miss Farrow in *Husbands And Wives*: "passive-aggressive." She insists her book isn't an act of revenge, but on the jacket blurb she's happy to license a sneer from her chum Gore Vidal: "A terse tell-all style of such candor that she even makes - inadvertently - her last husband, Philip Roth, into something he himself has failed to do - not for want of trying - interesting at last."

In some ways, the jacket tells its tale as much as anything inside. The two male friends she turns to for enthusiastic endorsements are Vidal and John Gielgud, neither of whom has any reputation - how shall we put this? - as a ladies' man. Inside, her personal odyssey has a kind of tragic neatness: she loses her virginity to Richard Burton; then she marries Rod Steiger, followed by Hillard Elkins (producer of *Oh, Calcutta!*), followed by Roth. Less a series of relationships than of disastrous star turns: the Womanizer, the Father Figure, the Sexual Sadist, the Mental Sadist. Professionally, too, she has come full circle - from Charlie Chaplin to Woody Allen, very different funnymen but with a shared preference for child-women. Given her experience with actors, authors, and producers, you can only hope she'll draw the reasonable conclusion and take up with a logger or truck-driver.

In fact, for all the fuss about her portrait of Roth, the crux of her argument comes in the chapter on her second husband. "Elkins' entire being was centered on sexual gratification," she writes. "His fantasies were alternately voyeuristic and sadistic. Inexperienced and sometimes apprehensive, I was a willing partner to his games stretching the boundaries of physical experience."

She drew on these "adventures" when she played Blanche in *A Streetcar Named Desire*, though not without reservation. "It is both shameful and courageous," she says, "to take a record from life and use it as a means to an end. The painter Claude Monet, to his own shame, looking at his adored young wife on her deathbed, could not help recording the changing color of her skin and the dissolution of her once-beautiful face. But he went on to use this image in his work."

That's really the heart of the book: artists make art from what they know, which means they cannibalize what, to anybody else, would be the most intense, private experiences. "Artistic" is a word loaded with associations: neophyte writer John-Boy Walton was "artistic," which, in TV shorthand, means uniquely compassionate, uniquely sensitive, uniquely blessed with insight and feeling. A third-rate British novelist (that narrows it down to a few thousand) once told me of an affair he was having with the wife of an insolvency lawyer: "We belong together!" he roared, and contemptuously dismissed his rival. "He has no soul!" So she left the lawyer, moved in with the soulful artist, and, after six months of mental and physical cruelty, fled gratefully back to the non-soulful attorney.

Another writer (there's no shortage of them) likes to try it on with chicks with the line, "As a writer, I'm sensitive to the needs of women." Amazingly, this line is so effective that, as he occasionally complains, it's dramatically cut into the time he has available to write. Would it work for anybody else? "As a plumber, I'm sensitive to the needs of women"? As an accountant? As a realtor? Yet you could easily make the case - with supporting statistics and graphs, too - that the practitioners of these crafts and many more have more understanding of "a woman's needs." Far from being possessed of especial sensitivity and compassion, the distinguishing feature of the great thinker's approach to the opposite sex is self-preoccupation and immaturity. "A Jewish man with parents alive is a 15-year old boy," wrote Roth in *Portnoy's Complaint*, "and will remain a 15-year old boy till they die." But Roth's have died, and he's still a 15-year old boy: for all the talk about his depression and inner demons and unwillingness to have his creativity constrained by domesticity, in the end he left Miss Bloom for another woman - a younger woman, and a friend of hers, too.

And don't look for mature reflections on the fairer sex from Chaplin or Allen or any other artist afflicted with the Pygmalion complex. My old friend Alan Jay Lerner had eight wives (the only Broadway record Andrew Lloyd Webber hasn't beaten), which seems ridiculous until you look at his work: *My Fair Lady* - older, worldly man takes in hand young unsophisticated girl he thinks he can mold; *Gigi* - older, worldly man takes in hand young unsophisticated girl he thinks he can mold; and, in the Seventies, he musicalized *Lolita* - older, worldly man takes in hand... Alas for Alan, that was one reprise too many of "Thank Heaven For Little Girls."

To be sure, there are no shortage of takers for the part. "I could be his Muse, if only he'd let me," sighs a typical Roth female character in the epigraph to his *My Life As A Man*. For "Muse," read "source material." While dreary executives fret about not wanting to take their work home with them, the creative soul faces the opposite problem:

the eternal temptation to take his home to work. He gets a great novel out of it, but at what cost: to judge from Bloom's account, the home life of such a self-plundering writer eventually leads to complete loss of real feeling, its gradual extinction by artifice and effect.

In the closing pages of *Leaving A Doll's House*, the great man, having devastated his ex-wife and reduced her to a one-room rental in New York, sends her a note: "Dear Claire, can we be friends?" They meet in a restaurant. They order coffee. There is a long silence. Then he launches into 20 minutes of impersonal, superficial banter. Eventually, she interrupts: "Philip, why do you want to be friends with me?"

A smile teases his lips: "Oh, perversion…"

Does it sound familiar? The coffee, the silence, the banter, the teasing smile? We've read the scene in a hundred novels, including a few by Roth; we've seen it in a thousand movies, including a few with Miss Bloom. It is the final transformation of their relationship, from life to art. Philip has become "Philip" and Claire "Claire": he's not capable of being her ex-husband, only of *playing* him - auditioning the moment for some great work as yet unwritten. Marriage is ultimately only the research experiment for the art, and, if in the process of dissection, the poor laboratory mice die, well, it's in service to a greater cause.

I'm not being entirely metaphorical here. As writers go, Claire Bloom could have done worse than Roth, and, if it's any consolation, she was lucky to escape with purely mental torture. In 1994, George Steiner wrote, as one great thinker on another, "The thinker inhabits fictions of purity, of reasoned propositions as sharp as white light. Marriage is about roughage, bills, garbage disposal, and noise. There is something vulgar, almost absurd, in the notion of a Mrs Plato or a Mme Descartes, or of Wittgenstein on a honeymoon. Perhaps Louis Althusser was enacting a necessary axiom or logical proof when, on the morning of November 16, 1980, he throttled his wife."

Althusser is the noted French philosopher, though these days he's not noted at all for his philosophy but only for his resolution of the conflict between his calling and his domestic arrangements in that hotbed of French intellectualism, the École Normale. "I pressed my thumbs into the hollow at the top of her breastbone and then, still pressing, slowly moved them both, one to the left, the other to the right, up towards her ears where the flesh was hard," he wrote. "Helene's face was calm and motionless; her eyes were open and staring at the ceiling."

Take my wife. Please.

NECK

Tie

February 22nd 2001
The National Post

WHENEVER I meet *National Post* readers, they rarely comment on my insight and perception. Instead, they want to know why I'm not wearing a tie in the picture that accompanies this column. Andrew Coyne, Norman Doidge, Rod McQueen, you name 'em, they've all got ties, while the token open-necked bylines – Paul Wells, Noah Richler – at least look like a principled rejection of ties, their shirt collars flapping breezily across their shoulders like a cut-price charter stuck in a holding pattern over Mirabel. "But you," said one reader accusingly, "your picture looks as if you're *supposed* to be wearing a tie but it fell off on the way to the office."

Well, here's the reason. I lent my tie to Tony Blair. True story. This was a couple of years back, at the BBC's New York studios, where I happened to be early one Friday morning when a callow youth wearing jeans and a ghastly leisure shirt wandered in from Fifth Avenue. I assumed from the mad, random grin that he was recruiting for some sort of religious cult and was about to call security when he announced that he was Britain's Shadow Home Secretary and had been obliged to interrupt his Manhattan vacation to respond to that week's prison breakout "down the line", as they say, to BBC Radio News in London. He did such a good job that they asked him if he wouldn't mind saying the same things all over again, this time for the TV news. They could film him above the waist, so the jeans wouldn't show, but he was still concerned about his open neck and, as I was the only guy in the building, he inveigled me into handing over my tie – a tasteful yet splashy number from Ogilvy's in Montreal, if memory serves.

A couple of weeks later, I was in Britain and sought out a political insider: "Ever heard of a fellow called Tony Blair?" I said. "Claims to be Shadow Home Secretary. I lent him my tie."

"Are you nuts?" said my friend. "He's now the new Labour Party leader, 87 points ahead in the polls, the most popular man in the country after Tinky-Winky in the Teletubbies. He's completely revitalized the Labour Party, got rid of all that socialist baggage from the past. He wants to sever the party's union ties."

Well, I can't say I blame him, I thought. All that polyester and dull motifs.

I met Tony again a few days later when we were both on the David Frost morning show. At breakfast afterwards, he was churlish enough to deny me full credit for his victory in the leadership race, but he did concede, apropos that BBC interview, that back at Labour HQ no-one had paid any attention to anything he actually said but they did comment favourably on the tie. The spin doctors, not to say the knit and weave doctors, thought its bold colours projected both authority and innovation, gravitas and dynamism. You'll notice that my own neck now projects none of these qualities.

As the Bard said, some are born great, some achieve greatness, and some have great ties thrust upon them. Tony is in North America this week for talks with Messrs Chrétien and Bush, and flies in fresh from sending in the RAF to bomb Saddam – an Anglo-American operation that, according to Washington, was not Dubya's idea but Bomber Blair's. For all its faults, Britain is still the only other serious western country. London punches above its weight: its defence spending is, to all intents, as pitiful as Ottawa's, yet it's prepared to send its forces to bomb and kill in its national interest, as opposed to marketing them as international crossing guards, the wretched state to which our own historic regiments are now reduced. But aside from his strange appetite for warmongering – during the Kosovo business, while Clinton was under the desk consulting his polls, it was General Blair who wanted to push on and occupy Belgrade – the British Prime Minister, viewed from Canada, seems oddly familiar.

Since he took office in 1997, bewildered observers have tried to get a handle on what's known as "the Blair project". Taking advantage of the vast powers that accrue to the Queen's first minister under the Westminster system, Mr Blair has set about remaking the United Kingdom with an unprecedented zeal. For example, he's abolished the old hereditary House of Lords and replaced it with an all-appointed upper chamber of cronies and has-beens. He's introduced "asymmetrical federalism" to Scotland, Wales and Northern Ireland, allowing the last to maintain privileged relations with foreign governments (the Irish Republic), as Quebec does with France. An Ottawa-style Upper House of pliant deadbeats, "distinct society" status for problematic territories, the introduction to Scotland of a local Parliament divided between separatists and "soft nationalists", the unceasing demand that ancient institutions and symbols "modernize" themselves for the needs of a multicultural society... After four years, it seems pretty clear that what Tone's exciting "New Britain" boils down to in practice is boring old Canada. "Cool Britannia" (as the Blair regime dubbed itself) is cool mainly in the sense that Nunavut in February is. One assumes that New Labour is remodelling the mother country along the lines of its forgotten lion cub mostly unintentionally, but nonetheless, when a Blairite think-tank proposes replacing the Union flag with some designer logo, they are, consciously or not, searching for a British Maple Leaf.

In 1997, no Fleet Street analyst foresaw that Mr Blair would embark on the abolition of Britain and its substitution by a new nation made in Tony's image, as Trudeau remade us in his. Where did it come from? One could certainly argue that there's something quintessentially Canadian in the Prime Minister's smug niceness. But I think back to our encounter in New York and find myself nagged by guilt: Did some sinister Canadian organism jump the evolutionary chain from my Ogilvy's necktie to Mr Blair's bloodstream? On the day I bought it, had the elderly M Trudeau perhaps tried it on? Did some fatal Trudeaupian virus seep through Mr Blair's shirt and lead him on to the Canadianization of Britain? Surely no sane person would deliberately model every single one of his constitutional reforms on Canada, given that Canada is one of the great constitutional swamps of the western world. Surely, if he was going to mimic North American federalism, he'd take up successful US ideas not failed Canadian ones. But the tie I loaned Tone is now a noose around the collective neck of the United Kingdom. Britain's second Canadian Prime Minister (after New Brunswick's Andrew Bonar Law, who briefly occupied Downing Street in 1922-23) is defining the country out of existence.

Perhaps, while he's here, Mr Blair will wear my tie – signalling, like Bill Clinton wearing Monica's tie on TV during the impeachment business, that he still cares. In those far-off salad days of the mid-Nineties, cocky Westminster Tories thought Tony was a lightweight, a chicken pretty boy in a fancy necktie. As another British Prime Minister remarked on another visit to Canada, "Some chicken, some neck."

OESOPHAGUS

Here's the beef

May 18th 1996
The Daily Telegraph

I N RESPONSE to Senator Dole's challenge – "Who would you trust your children with? Bob Dole or Bill Clinton?" - the President has raised the stakes: "Suppose," he said, trying to keep a straight face in a speech to White House correspondents, "you go home tonight and you decide to order a pizza. Who do you trust to select the toppings? Bob Dole or Bill Clinton?"

CNN and *Time* magazine took him at face value and commissioned a poll: 54 per cent of Americans said Clinton, 26 per cent said Dole, and 20 per cent were undecided (or possibly not hungry). They polled 1,011 adults, which means a sampling error of plus or minus 3.2 per cent - that's to say, two slices of pepperoni and an anchovy.

At first, I'd assumed the pizza was metaphorical. After all, in the high-cholesterol American vernacular, every other food is. The last time I interviewed Pat Buchanan, he was most exercised by the fat profits Mrs Clinton had made trading cattle futures. "The First Lady says it's chickenfeed, but that's a lot of apple sauce," he said, though he conceded she was one tough cookie. When the rhetoric's really cookin', American politics is an All-U-Can-Eat Nite at a Surf'n'Turf roadhouse - and, as Harry Truman might have said, if you can't stand the heat, stick to the salad bar.

In the 1984 election, Walter Mondale campaigned on the slogan, "Where's the beef?" In the 1994 Congressional elections, the Republicans campaigned on the principle that there was too much pork. The only non-culinary election in recent years was in 1988, when George Bush kept calling for "a thousand points of light". Nobody knew what he meant, until Ben and Jerry, Vermont's hippy-dippy ice-cream makers, figured that he'd misread the script and had meant to say "a thousand pints of Lite". So that's what they sent him.

It occurred to me that in Britain the badly trailing Tories could use a good food slogan, so I called up and tried out the Mondale line on a source close to the Prime Minister: "Where's the beef?" I said.

"In Chirac's bloodstream by now," he said, and gave a maniacal laugh.

So it was back to chew the fat with my contact in Washington, who tried to explain:

"Take Clinton and the minimum wage bill. It's full of pork, there's no beef, and that turkey's chicken. But I got bigger fish to fry. Are you ready for lunch yet?"

"So where does that leave Dole?"

"Well, that chowderhead's in a stew because Clinton's tossed him a hot potato. Yeah, he's still got plenty of lettuce, but Dole's toast. They'll make hamburger outta him."

"So Clinton will win?"

"That chicken-livered, candy-assed, cream-puff butterball? Sure! The big enchilada'll give us the usual waffle and we'll lap it up. It's a picnic, easy as pie, piece o' cake. Shall we order in?"

"Dole's vice-presidential nominee could make a big difference," I suggested.

"Baloney!" he said. "The second banana don't mean beans. He's chopped liver. Strictly small potatoes. He can be the icing on the cake but he can't bring home the bacon when you're in a pickle 'cause you got a limp noodle of a cold fish as the big cheese. C'mon, man, wake up and smell the coffee!"

I came away feeling I'd gleaned several important insights into the political scene, if only I knew what they were. In America, what with the ever greater number of speciality flavours, it's very difficult to smell the coffee. Indeed, it seems to me unlikely, when you order a raspberry-hazelnut decaf latte with vanilla sprinkles, that there could be any room left for the coffee.

It's hardly surprising that more than a third of the population is obese when the vernacular is so high in polyunsaturates. And a culture that describes every activity in terms of food tends to give the impression that, whatever it's doing, it'd prefer to be eating. Even sexual slang sounds as if, left to their own devices, the participants would rather be down at Dunkin' Donuts.

By the way, if you are down at Dunkin' Donuts, check out the new gal behind the counter.

"Is that some cupcake!" said Bud. "Great melons!"

"Great buns!" said Earl.

That didn't leave much of her anatomy for me to enthuse over, but I did my best: "Great hams?"

That's why the pizza line works for Clinton. It makes explicit what Americans have always sensed: this is a President who looks like the American language; he is the vernacular made flesh, the ultimate big cheese. Dole, by contrast, aside from a 1994 speech in which he told the Senate that "we need to de-pork the crime bill", is rhetorically undernourished. His idea of an aural zinger is "filibuster", which ought to mean two pounds of cream cheese with pickles and lettuce on a sesame seed bun but sadly doesn't.

Asked by a meeting of Republican women to name some recent achievements, he explained proudly that, on the health insurance bill, he'd managed to manoeuvre Senator Kennedy into a corner and force him to filibuster himself. The ladies looked alarmed: whatever "filibuster" means (and it sounds like a very literal definition of Ted Kennedy), the idea of forcing the Senator into a corner to do it to himself is not an appealing image.

Poor old Dole. In the Fifties, Eisenhower advanced the "domino principle" - the danger that developing countries would fall one by one to communism. In an era of micro-politics, Clinton has given us the Domino's principle - the President as the nation's pizza delivery boy. He promises the American people an extra-large deep-dish customised precisely to their tastes. In their heart of hearts, they know it'll arrive an hour late and be a small thin-crust, cold and congealing, and half the green peppers and mozzarella will have fallen off the back a couple of miles back. But, at the time, it sounds awfully good. Bob Dole should face up to reality - in politics, there's no such thing as a fat-free lunch.

SPINE

less

December 17th 2002
The Jerusalem Post

I WAS READING a Salman Rushdie column the other day and, not for the first time, agreeing with 95 per cent of it. In fact, I agree with him so often these days I've almost stopped noticing it.

But not quite. Far away at the back of my mind, I still remember the Rushdie of the 1980s - reflexively leftist, anti-Thatcher, the works. The old line – a neoconservative is a liberal who's been mugged - goes tenfold for him. He's not just a liberal mugged by reality; he's a liberal whom reality has spent the last 13 years trying to kill. I still have difficulties with his novels, not least the one that got him into all the trouble, but in his columns and essays he has outgrown his illusions.

At the time – Valentine's Day 1989 - most of us in Britain and the West didn't appreciate the significance of the event. It marked the first time the Ayatollah Khomeini had claimed explicitly extraterritorial authority. Why he chose an obscure and for most of us unreadable English novel for his expeditionary foray is unclear, but the results must have heartened him tremendously.

Rushdie had not set out to offend Muslims. None of the London reviewers found anything controversial in the book. When British Muslims and their co-religionists around the world burnt copies of *The Satanic Verses* in the streets, BBC arts bores held innumerable discussions on the awful "symbolism" of this assault on "ideas."

But it wasn't symbolic at all. They burned the book because nothing else was at hand. If his wife and kid had swung by, they'd have gladly burned them instead. Overseas, they made do with translators and publishers. Rushdie's precious lit. crit. crowd mostly opposed the fatwa on the grounds of artistic freedom rather than as a broader defence of Western pluralism. That was a mistake.

In the Fifties and Sixties, Nasserism attempted to import Soviet socialism to the Middle East. It never really took. A generation later, the Ayatollah came up with a better wheeze: Export Islamism to a culturally defeatist West. Everything that has become pathetically familiar to us since September 11th was present in the Rushdie affair.

First, the silence of the "moderate Muslims": A few Islamic scholars pointed out that the Ayatollah had no authority to issue the fatwa; they quickly shut up when the consequences of not doing so became apparent.

Second, the squeamishness of the establishment: Rushdie was infuriated when the Archbishop of Canterbury lapsed into root-cause mode. "I well understand the devout Muslims' reaction, wounded by what they hold most dear and would themselves die for," said His Grace.

Rushdie replied tersely, "There is only one person around here who is in any danger of dying."

Roy Hattersley, the Labour Party's deputy leader, attempted to split the difference by arguing that, while he of course supported freedom of speech, perhaps "in the interests of race relations" it would be better not to bring out a paperback edition. He was in favour of artistic freedom, but only in hard covers - and certainly, when it comes to soft spines, Lord Hattersley knows whereof he speaks.

Gerald Kaufman, a Jewish MP who has since gone on to disown Israel all but totally, attacked critics of British Muslims: "What I cannot accept is the implication that it is somehow anti-democratic and un-British for Mr Rushdie's writings to be the object of criticism on religious, as distinct from literary, grounds."

Kaufman said this a few days after large numbers of British Muslims had marched through English cities openly calling for Rushdie to be killed. In the last few months, several readers have e-mailed me with their memories of those marches. One man in

Bradford recalls asking a West Yorkshire police officer why the "Muslim community leaders" weren't being arrested for incitement to murder. The officer said they'd been told to play it cool. The cries for blood got more raucous. My correspondent asked his question again. The police officer told him to "fuck off, or I'll arrest you."

And, most important of all, the Rushdie affair should have taught us that there's nothing to negotiate. Mohammed Siddiqui wrote to *The Independent* from a Yorkshire mosque to endorse the fatwa by citing Sura 5 verses 33-34:

The punishment of those who wage war against God and His Apostle, and strive with might and main for mischief through the land, is execution, or crucifixion, or the cutting off of hands and feet from opposite sides, or exile from the land. That is their disgrace in this world, and a heavy punishment is theirs in the hereafter. Except for those who repent before they fall into your power. In that case know that God is oft-forgiving, most merciful.

Rushdie seems to have got the wrong end of the stick on this. He suddenly turned up on a Muslim radio station in West London one night and told his interviewer he'd converted to Islam. Marvellous religion, couldn't be happier, praise be to Allah and all that.

The Ayatollah said terrific, now you won't suffer such heavy punishment in the hereafter. But we're still gonna kill you.

As bad as the fatwa was, the inability of the establishment coherently to defend western values was worse. Clifford Longley, the Religious Affairs Correspondent of *The Times*, was one of the few to understand what was at stake. The British government must surely know, he wrote, that some Muslim beliefs, "at least at face value, are not compatible with a plural society: Islam does not know how to exist as a minority culture. For it is not just a set of private individual principles and beliefs. Islam is a social creed above all, a radically different way of organizing society as a whole."

Longley wanted anyone parading a "Death To Rushdie" placard to be "taken at his word and arrested for incitement to murder. The immediate consequences could be unpleasant, even including the risk of riot. But the painful shock of such a confrontation may regrettably be necessary before the British Muslim community is brought face to face with the reality that tolerance and compromise, even over fundamentals, are a fundamental requirement of life in Britain." Instead, all those British Muslims who called openly for Rushdie's death are still around, more powerful and with more followers.

Her Majesty's Government lacked the will then, as most of the West does today. In effect, the Ayatollah was allowed to get away with annexing Islam for political purposes, not just at home but internationally. If "moderate Muslims" are a viable demographic at all, they face a choice: They can follow the murder-inciters of Bradford, the suicide-bombers of the West Bank and the depraved killers of northern Nigeria on their descent into barbarism. Or they can wake up and save their religion. Either way, the West will be little use.

NERVOUS SYSTEM

Beyond words

February 4th 1999
The National Post

The director of D.C. Mayor Anthony A. Williams's constituent services office resigned after being accused of using a racial slur, the mayor's office said yesterday. David Howard, head of the Office of Public Advocate, said he used the word "niggardly" in a Jan. 15 conversation about funding with two employees. "I used the word 'niggardly' in reference to my administration of a fund," Howard said in a written statement yesterday. "Although the word, which is defined as miserly, does not have any racial connotations, I realize that staff members present were offended by the word." – The Washington Post

A S EVERYONE in the world knows by now, the head of the District of Columbia's Office of Public Advocate resigned last week for not using a racial epithet. And ever since the non-racial epithet in question – "niggardly" - has been making a spectacular comeback. The world's press has certainly not been niggardly in letting columnists niggle over "Niggardlygate", as it's known down here inside the Beltway. A Nexis/Nigxis search revealed 14,873 citations of the word "niggardly"' in the last five years, all but three of them from the last week.

"Niggardly", we're told, is nothing to do with "the n-word", but comes from "nig", meaning "miser" in Middle English and has older roots, according to the dictionary at hand, in Old English, Old Icelandic and Middle High German. The word's antiquity is apparently the mitigating circumstance. Those Middle High Germans,

Old Icelanders and Old Englishmen are so old they predate whitey's systemic racism against the black community: They were too busy being abusive to the Welsh and the Flemish to think up offensive words for African-Americans.

It was never very likely that this subtle explanation would cut much ice in the grievance culture of Washington. Unlike the White House, where "it depends what the meaning of the word 'is' is," in the municipal government it doesn't depend what the meaning of the word "niggardly" is. Either way, when you cross the District's borders, you're entering a Dictionary-Free Zone.

So, a week later, David Howard remains out of a job for telling his subordinates he'd "have to be niggardly" with the budget for the Constituent Services office. The African-American aide who stormed out and denounced his boss had been a rival contender for his job. The city's newly elected black mayor, who compared Mr Howard's conduct to "smoking in a refinery," had recently been criticized by *The Washington Post* for not being (as the headline bluntly put it) "black enough" - unlike his predecessor Marion Barry, the celebrated coke-snorter and wily race exploiter, who was black to a fault.

Most DC council members agree that Mr Howard had to go. "It certainly shows poor judgment to use a word that sounds like the n-word," said Ward 2 councillor Jack Evans.

Only the city's gay community is outraged: David Howard was the new mayor's first openly gay appointee - and after all, as one said to me, he hadn't used a bad word, only a word that sounded similar.

"You mean a homophone?" I said.

"Pardon me?" he said, no doubt wondering what the Gay Helpline had to do with it.

But Washington's campaign against, er, homophonia would seem to be strictly one-sided. If Mr Howard's offended black aide had said, "Prolonged budget negotiations will not be fruitful," it's unlikely his fruit of a boss would have flounced off in a huff. Jews, too, have learned to shrug off the references by the Reverend Jesse Jackson, the Fruit of Islam Minister Louis Farrakhan and other black leaders to "kikes" and "hymies": no Jewish entomologist would storm out because his superior described an insect as hymenopterous. Jews are not anti-semantic.

Elsewhere, though, identity politics is an increasingly fraught business. Last year in Houston, an African-American city official had to resign after calling a midget a dwarf, which isn't the same thing at all. A food critic for *The Dallas Morning News* wrote of one restaurant that its bland menu was due to "a niggardly hand with seasonings": the chef turned out to be black, and, after complaints, the word has now been banned from the paper.

But Niggardlygate is the first time a white guy's had to resign for disparaging himself in an inappropriate way. If there's a subtext to this determined obtuseness, it's that Washington is anti-niggardly in more than the linguistic sense: under the bloated administration of Mayor Barry and his incompetent cronies, city budgets bore no relation to fiscal reality. And, given that most of those cronies were African-American, if we are to ascribe racial characteristics to innocent words, you could certainly argue that, in the context of DC public servants, "niggardly" is an anti-black epithet.

Should Mr. Howard have had to quit? As everyone outside Washington agrees, no. But here's a trickier one: what's so wrong with the word "nigger"? I only ask the question because in the last ten years "the n-word" has been resurrected from the linguistic grave – by blacks. Young African-American men routinely call themselves "niggaz," form rap groups called Niggaz With Attitude and record songs called "Strictly For My Niggaz." Indeed, this fashionable spelling may be one reason for the etymological confusion over "niggardly."

Furthermore, most rap records are bought by white suburban teens, which suggests that the next generation of honkies will have a greater ease with the N-word than their parents - even if, given their fondness for the more robust forms of black culture, they'll presumably use it, as gangsta rappers do, as a term of endearment. It's simply not credible to demand that we segregate the dictionary - to say a Time-Warner executive can sign a hot rap act that sings "Strictly For My

Niggaz" but that, if he says he's going to have to be niggardly with the contract, he has to lose his job. Jews do not take kindly to being stereotyped as miserly, but nor do they go around forming hot new Yiddisher rap acts called Niggardly With Attitude.

You might think that a government employee who failed to recognize a word contained in every pocket dictionary would be shamed into silence rather than mounting a campaign to punish somebody else for his ignorance. But American "sensitivity training" has so effectively banished public expression of non-approved views that the only growth potential for the country's huge race industry lies in wilful misunderstanding. Not so long ago, for example, a Texaco executive was assailed for using "the n-word" in an illegally taped board meeting: a "transcript" of his words was read out on all three major US TV networks as evidence of the company's racist policies. When the original, poorly recorded tape was eventually released and digitally enhanced to improve the sound quality, it emerged that the guy had been making some remarks about Santa Claus and that neither "the n-word" nor any other racial slur had passed his lips.

Nonetheless, Jesse Jackson, the nation's Number One shakedown artist, succeeded in using the inaccurate transcripts to pressure the company into coughing up gazillions of dollars and instituting further "sensitivity" programs. It was a great victory for the forces of "tolerance" - the tolerance of stupidity, the dominant trend in American public life. And that's what Mr Howard's resignation shares with Bill Clinton's truly Orwellian assault on language over the past year: if you don't agree on what words mean, you're unlikely to agree on anything else. You can't have an integrated society without an integrated vocabulary. But, in Washington, DC, at both the federal and municipal level, it's no longer possible to call a spade a...

On second thoughts, let's not even get into that.

BACK

Where he belongs

August 25th 2001
The Spectator

ACCORDING to his tanned spokesman, George W Bush will cut short his vacation in Crawford, Texas, and return to Washington next Friday, August 31st. The President arrived in Crawford on August 4th and it was thought he intended to stay at least until Labor Day, September 3rd, thus beating Richard Nixon's 1969 summer sojourn and earning his place in history as the taker of the longest-ever presidential vacation. On the other hand, even at a paltry 28 days, it's almost certainly the longest vacation anyone's ever taken in the Greater Waco area. Don't try to book online: the computer will redirect you to more glamorous resorts such as Crawford, Florida, Crawfordsville, Indiana, or Crawford Notch, New Hampshire, and, if you insist that you really want to spend a month in Crawford, Tex, the entire site crashes.

Vacation-wise, Bush's place in history is already secure, as the patron of the hottest presidential resort in history: in his usual careless, brutal way, Dubya has ended the bipartisan presidential tradition of moderate vacation destinations with average August temperatures in the mid-70s - Clinton, Martha's Vineyard (77); Bush *père*, Kennebunkport (75); Reagan, Santa Barbara (75). In an average August, Crawford clocks in at 97 degrees. This summer, if anything, it's a little hotter, with temperatures not dipping below three digits until well after sundown. Needless to say, the town, like the President, is teetotal.

The White House press corps breezed in three weeks ago, and discovered that Crawford - a dusty crossroads in the middle of a drought-stricken, sun-broiled plain, population 690 - has five churches but not a single hotel. So they have to stay 25 miles away in Waco, where the Chamber of Commerce keeps harassing them to come on its media barge cruises on Lake Waco. No reporters showed up for the first media cruise. But a couple of days later the second cruise attracted two Washington journalists, who'd evidently made the mistake of using up all the local colour in their first piece. By now, I'll bet that boat is standing room only.

In Waco, by the way, the local colour consists of the charred remains of the Branch Davidian compound and the Dr Pepper Museum, a shrine to the popular non-alcoholic beverage invented in the town. In Crawford, the local colour's thinner on the ground: the school gym, seven miles from the Bush ranch, has been converted to a "Western White House media center", where reporters can pick up complimentary brown bags from the Brown Bag, Crawford's only gift shop, so called because any gifts purchased are put in brown paper bags. Inside each bag is a small Texas flag bearing the legend "Crawford, Texas: The Texas White House" plus a coupon entitling the bearer to a free scoop of Blue Bell Ice Cream, and a postcard of Crawford.

Otherwise, nothing much happens in the gym: by week two, the press were reduced to taking artistic shots of the switched-off microphone on the empty lectern behind the big "Western White House" sign, the sort of thing *The Independent*'s front page used to be very partial to back in Stephen Glover's day. Reporters interviewed each other about what they liked best about Crawford: "You don't have to pay to park your car," CBS correspondent Mark Knoller told *USA Today*. You barely have to pay to park your house: a three-bedroom, air-conditioned home in Crawford costs $30,000, but don't worry; if that sounds a bit steep, you can get a couple of acres and a double-wide trailer for about a fifth of the cost.

Meanwhile, back at the ranch, George and Laura are in relaxed, non-newsmaking mode.

Don't forget, these reporters are guys who've spent the last eight summers with Bill and Hill on the Vineyard and in the Hamptons. And, if you need to ask, "Which Vineyard? Whose Hamptons?", there's a rusting double-wide in Waco with your name on it. My favourite Clinton vacation photo was from the summer of '98: after his disastrous mea sorta culpa re Monica - the Slicker's worst ever three minutes on network TV - he flew up to the Vineyard to be comforted by all the beautiful people. At the airport, Carly Simon embraced him and the puffy-eyed Bubba gratefully returned the compliment, smothering her and pressing his pudgy fingers into the gorgeous tanned expanse of her back. I was once in a packed elevator with Carly in a backless dress - Carly, I mean, not me - and I can tell you she has one of the all-time great backs - fabulous shoulder-blades, the small of her back glistening with the faintest perspiration like a shimmering desert mirage you long to dive into.

Where was I? Oh, yeah, Clinton. Anyway, the thing is, if you're a White House reporter and you're used to hanging out with Bill and Hill and Carly and Steven Spielberg and Alec Baldwin, it's something of a culture shock to have to spend a month trying to wring a quote out of Bill Sparkman, who's been Crawford's barber for 41 years. So unsurprisingly the media has moved on to musing, "Why would Bush do this to us?"

The obvious answer, given the general tenor of his press cuttings, is: why wouldn't he? But the media guys are working on the assumption that, like them, Bush would rather hang out with Alec Baldwin than Bill Sparkman and that therefore this Crawford business is part of some political calculation. In 1996, Bill Clinton famously got Dick Morris to take a poll on where he should go for a vacation, and wound up having a miserable time pretending to camp and hike and fish on the Snake River near the Grand Teton Mountains in Wyoming.

This was supposed to appeal to swing-voter soccer moms with kids who liked to camp out. True to Morris form, after the poll-driven vacation Clinton's numbers went down, and thereafter he stuck to kibitzing with Carly and co. Indeed, even Slick Willie at his slickest might have balked at the fine calibrations of this year's multiformat Tony Blair vacation (suffering a week in England before swanning off to the Continent). But, in such a world, Bush is naturally assumed to be doing the same thing.

"This is the anti-Clinton presidency, so you go to the anti-Clinton vacation spot," says Marshall Wittmann of the Hudson Institute. "Some analysts see careful image control behind his decision to eschew wealthy escapes such as Martha's Vineyard," reported Anne Kornblut in *The Boston Globe*.

"Maybe George W Bush is carrying this 'I'm-not-Bill-Clinton' thing too far," wrote Ronald Brownstein in *The Los Angeles Times*. "Spending a month in the Texas dust seems an overly emphatic recoil from Clinton's preference for chic vacations around the beautiful people." Bush, it seems, is only in Crawford as part of an ongoing White House campaign to boost his non-Washington image and connect him with rural, heartland values.

Now I like spin and cynicism as much as the next guy, but in this instance Brownstein, Wittmann and the gang are only spinning themselves. The idea that Bush bought 1,600 acres of Texas dust-bowl and then built a ranch on it just on the off-chance that, in the event he became President, it would piss off the media is, to say the least, a little far-fetched. Isn't it more likely that, incredible as it seems, he actually likes the town?

We can't say we weren't warned. During Campaign 2000, while Al Gore was running around hitting three states every 24 hours, Bush was back in Crawford, the first presidential candidate in history to spend election year working on his retirement home. At the end of August, the White House press corps gets to go back to Washington. But this is where George W Bush will be spending the rest of his life. Voluntarily.

"When you're from Texas and love Texas, this is where you come home," Bush told the press corps, as they rolled their eyes.

Yet having a home is what made him President. If Gore had won his alleged "home" state of Tennessee, he'd be in the White House now. But, of course, it wasn't his home, and since the election he's shown little inclination to make it one. Likewise, Bill Clinton, whose only home in Arkansas will be the penthouse apartment his official library is being fitted with (the first presidential library in the Republic to be so accessorised). Clinton's like one of those 1-900 phone-sex lines: he has no geographic area code; he's from everywhere and nowhere.

But in Crawford the ordinariness of Bush takes on an epic quality. I don't mean "ordinary" in a disparaging way. As anyone who's spent any time around celebrities will know, they can be just as dull as any Crawford hairdresser or short-order cook, the tedium only augmented by their obsession with status. And, if you ever get into big celeb parties, you're often struck by the knots of forlorn superstars making halfhearted attempts at stilted small-talk far more boring than anything you'd hear at the Elks' Lodge or the Ladies' Aid. Clinton, after all, does not hang out with great scientists or philosophers, but with Alec Baldwin. Is the notion that Bush might find Alec's company boring and unrewarding really so absurd?

Apparently so. And not just absurd but offensive to the media's insistence that the presidency should be a starring role. Fortunately for them there's always America's new Ex-President-For-Life. Marvelling at Bill Clinton's delirious reception in Harlem the other day, *The New York Times'* star columnist, Maureen Dowd, gushed that he was the "Sinatra of politics" - unlike his pygmy successor: "Just as W always seems smaller than his station and surroundings, Bill always seems bigger. One leaves no footprints and the other is all footprints."

Like Fay Wray in one of King Kong's craters, Ms Dowd seems to be having difficulty clambering out from Bill's massive footprint: "I've got you under my shin," as the Sinatra of politics would no doubt sing.

Bush, sniffs Ms Dowd, is "Eisenhower with hair". All hairless Ike did was win the Second World War; but he'd have looked like a dork trying to groove along with Stevie Wonder.

The Crawford summer seems to have been the final straw for Ms Dowd, raising the suspicion that across the fruited plain there might be far more Crawfords and Wacos than Harlems and Hamptons. Last Sunday she bemoaned that Bush's crabbed vision is leaving its non-existent footprints everywhere you look. "America has grown insular, isolationist, paranoid," she despaired. "Nothing leaps ahead. Power clings to the *passé*, retreating from the cutting edge… Everything - from Washington's trashed international treaties to the coal-and-drill Bush environmental policy to Hollywood's tedious remakes and endless parade of Second-World-War and Cinderella-themed movies - looks backward, not forward.

"Our missile shield, more science fiction than science, has become a metaphor for our passive, defensive, retro crouch."

Oh dear, oh dear. "Power clings to the *passé*" is actually the perfect summation of the average Clinton fundraiser guest-list: Spielberg, Streisand, Tom Hanks, Whoopi Goldberg… Not exactly "cutting edge", is it? Indeed, it's positively squaresville, like an especially lame issue of *People*. Come to that, isn't there something a little - dare one say - hicky in Clinton's need to cling to celebs and the media's need to cling to him?

We right-wing types are notoriously hard to please, and you won't be surprised to hear that policy-wise I've got grave doubts about this administration - the cave-in to Ted Kennedy on education, a zillion other things. But Bush the guy I like more and more. In fact, the man himself is at least as radical a project as missile defence: everything he does - or doesn't do - is a rebuke to the Clintonian notion that the role of the American people is to be the studio audience on "The *I'm The President!* Show". So instead of preening with "us" on the coast, he's out there with "them", plonked in the middle of flyover country. If Bush is, as Ms Dowd insists, "narrow, isolated and elitist", then I say bring it on, baby. It's like that old Fleet Street headline: "Fog

In Channel. Continent Cut Off." Dowdian Fog Off Nantucket. America Cut Off.

BONES

Lying in state

February 1998
The American Spectator

I WELL REMEMBER the first time I heard the name M Larry Lawrence. I was in the hot tub with the Princess of Wales when she said, "Is it true you were the real-life model for Al Gore?" and I sai…

Whoops! See how easy it is? And, after all, who among us has not occasionally told a little white lie on a résumé, a bank loan form, or a cemetery application? Still, in an age when everyone's lying, it takes some finesse to produce a world-class double whopper with cheese. In 1994, when Vice-President Gore swore in M Larry Lawrence as Ambassador to Switzerland, it must have been like the Stanley Cup of horse hockey:

Larry: "Thank you, Mr Vice-President. Speaking of foreign parts, did you know I'd been torpedoed off the coast of Murmansk?"

Al: "Really? Speaking of the war, did you know I was the model for Ensign Nellie Forbush in *South Pacific*?"

Larry: "Really? Speaking of *South Pacific*, did you know Rodgers and Hammerstein wrote it in my hotel?"

Al: "Really? Speaking of your hotel, did you know my third cousin died of second-hand smoke inhalation there?"

Enter the President.

Bill: "Mr Ambassador, I feel your coma. I have vivid and painful memories of convoys being torpedoed in the Arkansas of my boyhood."

Larry, Al, Bill: Liar, liar, pants on fire. As it happens, I too was a victim of M Larry Lawrence's tall tales. A few years ago, on the last night of a BBC TV shoot in San Diego, the local cameraman suggested we celebrate the end of the film by having a drink in the Hotel Del Coronado. (At the time of writing, Larry's ownership of the hotel is about the only item in his remorselessly shrinking curriculum vitae still standing. On the other hand, by the time you read this, it may well have emerged that even this was a fantasy and that he just wandered in off the street and started hanging around the lobby, dispensing bonhomie and asking if you wanted help with your bags.) Personally, I thought it was a ghastly place, but, lurking in one subterranean passage outside a karaoke bar from which some caterwauling no-talent was bellowing "I Will Always Love Yoolulating," I noticed a display of photographs of L Frank Baum and his *Wizard Of Oz* stories, with an accompanying note explaining that L Frank had written the original book while staying at M Larry's hotel. Not only that, but the chandeliers in the hotel's Crown Room had been designed by Baum himself. And Baum had based his design of the Emerald City on the Coronado. The director and I had long wanted to make a profile of Baum, and what better place than here? With an actor playing Baum, writing *Oz* up in his room while staring out at the ocean... The San Diego crew, anticipating another lucrative booking, agreed, and excitedly we all repaired to the bar.

We did a rough budget based on the Baum-in-San-Diego angle and then I flew home. Whereupon I discovered that not one contemporary account of *Oz*'s creation even mentioned the Coronado. It appeared, instead, that Baum had written *The Wonderful Wizard of Oz* while working as editor of *Chicago Show Window*, a trade magazine for window decorators. Strange, I mused, but never gave it another thought. The display of Baum working at the Coronado is still up there, notwithstanding that, as Larry's fulsome obituaries were succeeded by their emptysome rewrites, the *Oz* business, like Larry's Merchant Marine service and his professional football career and his vice-chairmanship of the Nobel Peace Prize Commission, proved to be the merest fictions. Somewhere in his youth, it seems, Ambassador Lawrence had misheard the Cowardly Lion's song:

Oh, I could show my prowess
Be a-lyin', not a mowess
If I Only Had Da Noive.

Larry had da noive alright. And it worked. More recent Baum biographers have been happy to play up the Coronado angle, adding credence, en passant, to the hotel's other fancies:

When it was built in 1887, the Coronado had more lights than any structure outside New York City. Thomas A Edison himself had supervised their installation.

Er, no, he didn't. That bit was Larry's idea. If, as the English diplomat Henry Wotton put it, an ambassador is a man sent abroad to lie for his country, M Larry Lawrence had been in training for his post all his life. Burial at Arlington hardly seems enough: he should have had a full-scale lying in state.

Instead, the poor fellow has been furtively disinterred from the National Cemetery and returned to San Diego. Let lying dogs sleep, I say. He should have remained at Arlington, with slashes chiseled through the relevant parts of his tombstone ("S1C – U.S. MERCHANT MARINE") in the same way that, in Budapest, the streets' pre-Communist names have been restored but the signs honoring Marx and Lenin have been left in place with red lines scored through them: lest future generations forget. Given the ease with which he bamboozled the White House and the State Department's "background checkers" ("bamboozled" is the official explanation, though it's more like a strange over-eagerness to look the other way), Larry Lawrence should have stayed in Arlington as a monument to the Clinton administration, a Tomb of the Unknown Donor: "But we didn't know he was a phony/a foreign national/an agent of Red China/an international drugs dealer/an Indonesian gardener making $15,000 a year (delete as applicable)."

Alas, Larry's been hustled back to San Diego in a laundry basket, his place on the nightly news already supplanted by El Niño or the new fat pill. Who cares if the only military record he had was "In The Navy" by the Village People? Larry pretended to be a war hero; Bob Dole actually was one - but, according to polls conducted a couple of months before the Presidential election, over 50 per cent of Americans had no idea he'd been in the army. In a present-tense culture whose collective memory of anything before the death of Elvis is increasingly wobbly, all that counts is the story you tell this week. Bill Clinton said recently that he hadn't eaten in McDonald's since becoming President. They've got video footage of him stepping through those Golden Arches, but so what? Likewise, the Vice-President, who, on successive days after his claim to have inspired Erich Segal's *Love Story* was exposed as hogwash, first stuck by his story, next said that journalists had "misheard" him, then denounced them for violating an off-the-record briefing. As Erich Segal almost wrote, Gore means never having to say you're sorry.

There are still places where truth matters. In the "Profumo scandal," which rocked Britain in the Sixties, it's worth noting that John Profumo was forced to resign as Secretary of State for War not because he'd cavorted with call girls but because he lied to the House of Commons; it wasn't the sex that ended his career, but the way the sex had rotted his integrity. Even in an age of sound-bite politics, Britain retains a sense that in public life this is the greatest sin of all. Thus, the author of a recent biography of Jonathan Aitken, a more recently disgraced Cabinet minister, chose for his title not a reference to Aitken's dodgy Arab business deals or his taste for sado-masochistic sex, but a blunt summation of his principal offense: *The Liar*. In modern British politics, there are only a handful to whom that title could apply. In America...?

Perhaps we shouldn't be surprised. So many areas of public debate in America are now conducted in such a blur of evasions that anyone with a functioning sense of integrity would instinctively recoil from such a world. Race is an obvious example, as President Clinton's woozy "national conversation" is proving all too well. But so are the hot-button issues du jour, "gay rights" and "global warming."

It comes as some surprise, then, that the guy who's championed all these bogus issues, Al Gore, is proving so spectacularly incompetent at being bogus himself, tying himself in knots over *Love Story*. Perhaps in claiming Tipper as the inspiration for the doomed Ali McGraw character, the vice president was just indulging his well-known taste for deathbed scenes: after his '96 Convention speech about the sister who died of lung cancer and the '92 Convention speech about the son who was nearly killed in a car crash, maybe he was worried about running out of stricken relatives. Who can say? What's clear, though, is that, even in a liars' culture, Gore is having a harder time of it than Lawrence and Clinton. It's not just that they're congenital liars, they're congenial liars. Lying comes naturally, effortlessly to them; hell, they enjoy it so much they do it even when there's no need, just for the exercise.

Gore can't manage that. Even his shiftiness is wooden. All of us at some point in our lives confront the George Washington moment in our own particular way. The President, for example, has his own unique form of mea culpa: "I cannot tell a lie. Cherry trees were cut down. That's why I'm challenging the Republicans in Congress to support the Cherry Tree Cutting Reform Bill I've proposed." Gore's equivalent came after *The Washington Post* caught up with him after the dead sister speech and asked him why, if he'd pledged himself to fight smoking after his sister's death in 1984, he was still claiming to be a proud tobacco farmer in 1988 and still taking tobacco money until 1990. His tortuous answer foreshadowed the problems he's had since:

> *I felt the numbness that prevented me from integrating into all aspects of my life the implications of what that tragedy really meant. We are in the midst of a profound shift in the way we approach issues. I really do believe that in our politics and in our personal lives, we are seeing an effort to integrate our emotional lives in a more balanced fashion.*

Oh, yeah? As M Larry Lawrence would say, tell it to the Merchant Marine.

SHOULDER BLADES

Revenge of the killer Dairy Queen!

May 27th 2001
The Chicago Sun-Times

Burlington, Vermont

"JIM'S A ROCK star now!" raved one local politician of the decaf-latte persuasion as Senator Jeffords (R. - wait a minute, D. - no, for the moment, allegedly I-Vt) brushed past and a cheering throng swept us into the packed lobby of the Radisson Hotel (ah, the charms of small-town Vermont country inns). Jim, who normally looks as if someone's twisting a pineapple up his bottom, seemed eerily relaxed, enjoying his new-found eminence as the world's most famous obscure senator.

But I don't think he's a rock star. He's more Peter Tork from the Monkees, if you can imagine Peter flouncing off in a huff and joining the Partridge Family. Just over a week ago, Jim Jeffords was an amiable goof, whose three-decade "Republican" voting record read like a guy who's holding the road map upside down - he voted against Reagan's tax cut but for Hillary's health plan, against Clarence Thomas but for partial-birth abortion. This is what we in the media call "a force for moderation".

But it took a most immoderate act to secure Jim his place in history. Hitherto, his greatest accomplishment has been his loyalty to the North-East Dairy Compact, an agribusiness boondoggle that ensures single moms and the like have to pay more for their children's milk, but which unaccountably gets a great press from the national media, who seem to think it's something to do with preserving all those picturesque weathered dairy barns they drive past on the way to their Vermont weekend retreats. At any rate, the Vermont Dairy Queen with the soft-serve principles has finally blown his cone: he's quit his party, and thereby ended the GOP's hold on America's longest continuously held Senate seat - Republican for 140 years. Better yet, he's brought a dash of Westminster horse-trading, a touch of Italian coalition politics to Washington: for the first time in US history, control of the Senate is passing from one party to another without anything so tiresome as an election.

The constitutional propriety of this has mostly gone unremarked. In Burlington, a leathery old plaid-clad lesbian lectured me about Bush's "illegitimacy" and the Supreme Court's "post-election coup." But, if it's wrong to install Dubya in the White House through one vote from an "ideological" judge, surely it's wrong to install Tom Daschle in the Senate Majority office through one vote from a senator peeved because Bush didn't invite him to the White House "Teacher Of The Year" reception, even though the winning teacher was a Vermonter.

Did I fall asleep and miss a constitutional amendment? Or has this rule been around since 1787? "Any sitting senator who findeth himself excluded from ye presidential receptions such as, but not limited to, Teacher Of The Year, Powdered Wigmaker Of The Year and Buggy-Whip Manufacturer Of The Year, hath the right to remove all officers of the Senate save himself from their posts." On such minutiae do empires rise and fall. Who knows? When Gavril Princip assassinated the Archduke Franz Ferdinand in 1914 and plunged the world into war, maybe he was steamed at not getting an invite to the Sarajevo Teacher Of The Year All-U-Can-Eat Buffet.

The press has roundly castigated Bush for his "meanness" and "pettiness"' over the Teacher Of The Year guest-list, and they may have a point, though not the one they think they have. For my part, I only wish the right were as tough as the other crowd. Last week, before Jeffords flew the coop, the Democrats were keeping the oldest

Republican senator, Strom Thurmond, on the floor hour after hour in one frivolous roll call after another, declining to let him "pair" with a Democrat and so retire early. The genial old sex fiend is 98 and as hot for the gals as ever. But ever since the election the media have been running a ghoulish Strom deathwatch: all it would take is a particularly nubile intern to come jogging bra-less round the Capitol and the 50–50 Senate would belong to the Democrats. Last week, as they put the ancient reformed segregationist through yet more 18-hour days of pointless procedural mischief, it was as if Minority Leader Tom Daschle and his troops had decided they'd waited long enough for ol' Strom to kick off, and it was time to hasten the process. On the Monday, some of the old boy's Republican colleagues were worried that he wouldn't last the night. You gotta hand it to those Dems: there's a party that knows how to play hardball. They don't just tear up your Teacher Of The Year invite, they measure you up for the Casket Of The Year competition.

No one's taken a keener interest in Strom's health than Jim Jeffords. The November election had left one otherwise unimportant man a window of opportunity, which wouldn't last for ever: Jeffords figured that, if Strom did keel over, Daschle would take charge of the Senate and owe Jim nothing; he'd be just another out-of-work GOP committee chairman. The Republican establishment in Washington claims not to have been aware until last Tuesday that Jeffords was checking out, which is very probably true given the general doziness with which Trent Lott and co have presided over the Senate. On the other hand, my friend Tom, who's currently painting my house and goes drinking with a tattooist who's well-connected in Vermont Republican circles, told me three months ago that Jeffords was planning to quit the GOP. That sounds more like it.

It's not strictly true that until last week Jim was entirely unknown outside Vermont. At my place in New Hampshire, the only TV station I can get is Channel 3 from Burlington, so I never hear anything about the Granite State's all-Republican Congressional delegation, but night after night the local news is full of Jim — Jim with dairy farmers, Jim with schoolkids, Jim announcing he's secured another X billion dollars for some idiot Federal program. Even if you had no idea that he belonged to the GOP's "moderate" wing, his campaign ads always suggested a certain sheepishness about his party: he was "proud," he told Vermonters during November's campaign, to have received the support of so many "Independents, Democrats and Republicans" — this last word mumbled sotto voce, like a schoolboy asking the pharmacist for condoms. You run into him everywhere in Vermont — county fairs, that sort of thing — everywhere, that is, except Republican Party events, which he pretty much stopped going to because he always got booed.

Indeed, the one good thing about his belated formal abandonment is that Vermonters can no longer cite their Congressional delegation as a perfect embodiment of the state's "diversity": one Republican senator (Jim), one Democratic senator (Patrick Leahy) and one Independent Socialist congressman (Bernie Sanders). In practice, this theoretical "diversity" resulted in a remarkable homogeneity: all three vote pretty much the same way on pretty much everything — that's to say, with the Democrats. Nonetheless, in his speech at the Radisson, after noting that Vermont was the first state to abolish slavery and that its per-capita death toll in the Civil War was higher than any other state, Jim couldn't resist claiming to be acting in the same long tradition of "principled" "independence" by his courageous act of stiffing his party and going "independent" after first taking the precaution of ensuring that the Democrats would reserve a powerful committee chairmanship for him.

By the way, I trust I wasn't being "homophobic" in my characterization of that leathery lesbian. Vermont is the first state in the nation to have legalized a form of "gay marriage," and, in Burlington, Jim's *hasta la vista* to the GOP brought out several female "civil union" couples, waving "WE'RE SO PROUD OF YOU, JIM" signs. I had a very pleasant time with two perky young Sapphists who yelped and high-fived every time Jim's more-in-sorrow-than-in-anger routine touched on the inadequacies of Republicans. It was a nifty idea to come home to abandon ship, and no doubt the pictures looked terrific on TV. But it wasn't what you'd call a typical Vermont crowd. If he'd given his speech in the less Ben & Jerrified *quartiers* of the state such as Newport or St Johnsbury, he'd have got a rougher ride. And, even in Burlington, Jim wasn't taking any chances. "He came here to talk to real Vermonters, but we're not allowed in!" yelled one female dissident in those broad North Country vowels you hear less and less, as the doors closed on the Senator's no-public-admittance press conference. He'd flown from Washington to Burlington so he could announce his defection in front of a home-town crowd of ABC, CBS, NBC and CNN crews.

In Britain, they talk of colonial administrators on remote islands "going native." Jim Jeffords has made the journey in reverse: he's a Vermont native who's gone flatlander ("flatlander" being the preferred term for incomers from points south). The old-time Yankee virtues that enabled his forebears to carve out a home in these hills 200 years ago were long ago abandoned by Jeffords: he favors the federalization of education, big-time entitlements, a heavy regulatory hand on almost everything. The pundits say, ah, well, but that just demonstrates how in tune Jim is with the new, "liberal" Vermont. But even Vermont isn't *that* liberal. In 1980 and 1984, Vermonters voted for Reagan, which you'd think might have stiffened even a jellyfish like Jeffords into supporting his President's budget. In 1998, a strong conservative pro-life gubernatorial candidate came within a whisker of deposing a popular Democratic incumbent and got 42 per cent of the vote, which is a better result than most British Tories can expect to see within their lifetimes. And just last year the Vermont GOP won the State House - and not wussy, "moderate," "tolerant" Republicans either, but cranky, angry white male-type Republicans steamed about gay marriage and logging rights (they're unconnected controversies, I hasten to add: you don't have to have a gay spouse in order to log, even in Vermont).

Not that their victory owed anything to Jeffords. He "declined to endorse" Republican State House candidates, State Senate candidates, the gubernatorial candidate or the Congressional candidate. Even so, his Democratic opponent called on him to distance himself from the divisive rhetoric of others in his party. Jim was flummoxed. Distance himself? If he were to distance himself any further, he'd be campaigning from Bermuda.

That the least surprising self-outing since George Michael declared he was gay should cause such havoc is principally the fault of Republican leaders going back 15 years. Not for the first time, the GOP's Senate backslappers called it wrong, and the fellows on the ground got it right. On April 24th 1984, the Republican Town Caucus of Kirby, Vt (population 347) unanimously adopted the following resolution:

> *Whereas Congressman James M. Jeffords has compiled a voting record of the sort one would expect from a fellow who can't pour maple sap out of a boot, even with the instructions printed on the heel,' they began, 'therefore be it resolved by the Kirby Town Caucus that the true Republicans of this town would cross hell on a rotten rail before they would vote for him again.*

But clubby Washington knew better. In 1988, when Congressman Jeffords decided to run for the Senate, Bob Dole and stiff-necked Mormon Orrin Hatch endorsed him in the primary, even though it was already clear that, whatever his other charms, Jeffords was no Republican, and never would be. And, as no incumbent senator has ever been defeated in Vermont, that's all the more reason for not giving the seat to an obvious time bomb. LBJ used to say of J Edgar Hoover and others that it was better to have them inside the tent pissing out than outside pissing in. But the Senate Republicans let Jeffords in the tent and he still wound up pissing all over them. Last year, the party faithful in Vermont had had enough and there was talk of a primary challenge. The GOP leadership, and Jeffords' aides, saw it off by assuring disgusted party volunteers that no matter how offensive Jim's votes were — he voted with Clinton 80 per cent of the time — the only vote that mattered was the one he cast to keep the

Republicans in the Senate leadership. In the last of many disservices to the Vermont GOP, Jeffords last week nullified that vote, too. "A POLITICIAN WITH A CONSCIENCE," read one sign in Burlington. "HONOR IS NOT DEAD," said another. A vain pliable boob who repudiated even his last residual pledge to his party for the most frivolous reasons is hailed as a giant of political integrity.

The Republicans are now deep in recriminations over "Who lost Jeffords?" But the real question is: "Who cares?" As Lott, Hatch et al have demonstrated, the GOP isn't cut out to run the Senate so it's for the best they no longer have to pretend they can. We'll see how deftly Tom Daschle manages the transition from Senate Obstruction Leader to Senate Majority Leader. Bush now has someone to blame, which could work for him in next year's elections. And, even if it doesn't, after all the billions they blew trying to keep Jeffords nominally in line over the years, surely even those dopey Republican Senate big shots must have learned an important lesson about letting the Trojan horse hang around to become one of the biggest nags in the stable.

SHOULDER

Pulled over

March 31st 2000
The Daily Telegraph

EVER SINCE Bill Clinton managed to wriggle free from the chains of impeachment, the experts have been saying this year's elections would be about "character". Well, the character issue finally blew into town and, before you could say Jack Robinson, it was, inevitably, the Republican it dropped a bucket of mud on.

As luck would have it, this Republican is Jack Robinson - a wealthy businessman who's running for Senate from Massachusetts. Last year, with the Clinton administration busy digging dirt on its opponents while publicly deploring "the politics of personal destruction", George Dubya Bush hired a private detective to dig up dirt on himself, just in case there were any heavy nights back in the Seventies he'd forgotten about.

Jack E Robinson III did the same thing, but, unlike Dubya, published the results in a *Complete Report To The Citizens Of The Commonwealth Of Massachusetts*. Mr Robinson revealed that he'd been arrested for an unpaid parking ticket; he'd plagiarised a book on the history of Pan Am; and a girlfriend had once taken out a restraining order on him.

The poor chump seemed to think he'd be treated as a Massachusetts McCain, hailed for his honesty and authenticity. Instead, he seems to have practised the politics of personal self-destruction. With his party in a state of shock, another young lady came forward to say that on a blind date he'd forced a kiss on her: "Jack the Tongue", she called him, and you could almost hear the ping of lightbulbs in the heads of the state's headline writers. The

147

Republican Governor, Paul Cellucci, abandoned him and the rest of the lily-livered GOP establishment went into hiding.

While trying to explain these developments during a live WBUR radio interview on his carphone, Jack the Tongue was struck by an out-of-control vehicle. "Boy, everything is happening to me!" he said. "Cellucci is withdrawing his support and people are sliding across the highway at me." Instead of immediately easing the Tonguemobile onto the shoulder and reporting the incident, he stayed on the road for a hundred yards until it was safe to pull over. But at least one eyewitness interpreted this as leaving the scene of an accident.

The next morning's *Boston Globe* neatly summarised the grim downward spiral of the Robinson 2000 campaign:

> *Jack E Robinson III, the US Senate candidate accused of sexual misconduct, plagiarism and carrying a dangerous weapon, found himself in a new controversy yesterday: defending his behaviour following a car accident.*

It all comes back to the character question, and the question is this: How much character do you need when you're running against the senior Senator from Massachusetts, Edward M Kennedy? How problematic is one overly probing tongue incident when you're running against ol' yo-yo pants? How speedily do you have to call in an accident when you're running against a guy who fishes himself out of the briny, goes back to his pad, and then remembers late the following morning, oh, yeah, there's some broad still down there?

Well, here's to you, Mr Robinson. You may be crazy as a loon, but so what? Any Republican willing to stick around Massachusetts is bound to be nuts. And on the other stuff you've got explanations: The restraining order was taken out by a gal who was, in fact, harassing him; the tongue complaint is anonymous. "I have never, at least as far as I know, acted inappropriately with any women," says Jack, and, although that qualification isn't the most reassuring, it's further than Teddy's prepared to go.

But the GOP have fled in terror: for the first time in 38 years the party won't be fielding a candidate against Kennedy, though Jack the Tongue says he's determined to run as an independent. He doesn't seem to realise what's happened to him, and I can't blame him: when you run against Ted Kennedy, you don't expect to be forced out of the race because of sex and driving.

After all, this is Massachusetts – or, as WRKO's Howie Carr calls it, Sodom and Begorrah. Unlike Senator Kennedy, Jack the Tongue didn't leave his date on the river bed. Unlike the Senator in less catastrophic mode, he didn't chase some hapless trolley dolly into the airplane washroom from where fellow passengers had to rescue her from an involuntary conscription into the mile-high club. Unlike Congressman Joseph P Kennedy III, his driving didn't cripple a girl for life. Unlike Congressman Gerry Studds, he didn't have sex with a 17-year-old Congressional page (male). Unlike Congressman Barney Frank, he didn't set up house with a rent boy known as Hot Bottom and fail to notice the cute little feller was running a thriving prostitution racket on the premises.

For decades many of us have wondered just what you'd have to do to be unfit for public office in Massachusetts. And now we know: plant an unwanted kiss and fail to pull off the highway on to the shoulder swiftly enough.

But, of course, all those other elected officials are Democrats. When Bill Clinton managed to shrug off oral sex with subordinates, harassment charges and credible rape accusations, we were told things would never be the same. And that may be true - for Democrats. But for Republicans it's as if Clinton never happened. As a successful African-American, Jack the Tongue is doubtless aware of certain police departments' habit of pulling people over for DWB – "driving while black". Unlike Senator Kennedy at Chappaquiddick, Mr Robinson committed an offence Massachusetts takes far more seriously: DWR - driving while Republican.

ARMPIT

Starting to sweat

September 21st 2002
The Daily Telegraph

Following the best-selling Zabibah And The King, The Impregnable Castle, *and* Men And A City, *we're proud to present an exclusive sneak preview of Saddam Hussein's fourth great allegorical romance!*

Saddam is the winner of the 2002 Governor-General's Award for Fiction for last week's reply to the United Nations ("We hereby declare before you that Iraq is clear of all nuclear, chemical and biological weapons"). An accomplished wordsmith in the tradition of Sheikh Spear, George bin Ard Shaw and Louisa May al-Cott, Saddam has given us exclusive rights to this excerpt from his forthcoming novel.

(As in his previous allegorical romance, the pliant female Zabibah represents Iraq and the stern King represents everyone's favourite dictator.)

LAST CAMEL FROM BAGHDAD
The latest great allegorical novel by Saddam Hussein

FIRST DRAFT

"Aaaar-oooow! Aaaieeeeeeeeee! Neeeeeaauuueaauuuu!"

"Zabibah, baby, can you hold it down? The neighbours'll think I've got a couple of Kurds clamped up to the electrodes again."

The nubile Babylonian maiden rolled back on the pillow, shook her ebony tresses over her lush full breasts and gazed devotedly at her king. "You are a superb lover, Majesty," she said. "The beads of sweat from your royal armpits are like nectar. The brush of your nasal hair is like cashmere. The stylish English trilby makes you look a little like Frank Sinatra on that LP you bring over when you're not in such a hurry and the power for the Dansette is working…"

"What was that?" said the King nervously, jumping up and grabbing his pants.

"What was what, O merciless master?"

"That faint whirring sound overhead."

"Oh, that's just the new electric fan. Come back to bed, my savage sovereign. Unzip your merciless trousers. You're in the no-fly zone, remember?"

"You're right," he said, slapping her around a little. "A woman is like a country. She needs a strong master to bend her to his will. You should be honoured that I do to you twice a week what I've been doing to Iraq for decades."

"I had that dream again last night, master," said Zabibah, peeling a green fig and playfully inserting it between the King's cruel lips. "The one where I seem to be the very landscape. My proud breasts are the northern hills, my gently undulating belly is the desert, my legs are the opposing shores of the Persian Gulf, and the eternal mystery of my womanhood is the great port at Basra. And then I look up and see a strange metal bird looming over me. I feel his shadow on my hills and desert and then his undercarriage opens up and he unleashes his enormous missile to penetrate me to the very core of my being. Then I wake up drenched in perspiration. What can it mean, my lord?"

"It means American men are lousy lovers," said the King. "They always explode on contact." He spat the fig from his mouth and almost took her eye out.

"You know this allegory business?" she said. "You're supposed to be Saddam and I'm supposed to be Iraq? What if that's not right? What if I, the helpless, slightly overweight woman, am really Saddam and you, the mighty warrior, are really George W Bush? Ever thought of that, big boy?"

The King was getting that weird headache again. And anyway how come the electric fan was working? The Americans bombed the power plant last Tuesday. And Zabibah's theory sounded awfully like that e-mail he got from Oliver Stone about the film version. Where was it again?

"Here it is," said Zabibah, pulling it from among the rose petals floating in the shimmering ornamental pool. "I'll read it to you:

From: stoner§paranoidpictures.com
To: saddam109§aol.com

Allegory, man! I LOVE IT!! What a book! I just read part of it all the way through. You're Zabibah, right? You've been taking it from the Pentagon, the CIA, Reagan all these years? And waddaya got to show for it? See, here's how we'll open. It's 1883, 1897, whatever, and Queen Victoria and Grover Cleveland are plotting how to control Mesopotamia for the entire 20th century, right? And first they figure, hey, let's change the name to Iraq, cuz then it'll sound like Iran, and no one'll give a f--- what we do to it. And they're totally out of their f---in' heads on f---in' opium and they go into this heavy trip where they see Dick Cheney shooting up a liquor store in Baghdad, and then he drops his pants and he's got, like, this huge gas pump in there and he fills up his Cadillac with it, and he's played by Gary Oldman, and you're like his bitch and you're whimpering, please, look, I did everything you wanted, I invaded Iran for you, please don't hurt me more than you have to, and Queen Victoria and Grover F---in' Cleveland are doing "Walk Like An Egyptian" by the Bangles.

"Stop it!" yelled the King, covering his ears and shattering the night silence with a piercing scream. In the hush that followed, he could just make out the loyal members of his faithful praetorian guard in the corridor saying, "Hear that? Someone's put a knife in the old bastard" and breaking into a chorus of "Ding Dong, The Witch Is Dead". "Look, Zabibah," he said, calmly. "For the last time, this is how the allegory works: I'm Saddam and you're Scott Ritter. No, wait…"

"Oh, it's all coming out now," she sneered.

"Shut up," snapped the King. "You might be a right purdy l'il honey lyin' there like - hang on, why am I suddenly talking in a Texan accent?" His headache was getting worse. This first draft was going nowhere. Maybe he should get some sleep, make a fresh start on Chapter One in the morning.

"Suit yourself," said Zabibah. "Maybe you're right. I'm a woman so I should be Iraq." She stretched lazily on the cushions and pulled a Turkish cigarette from his hatband. "But I'm beginning to feel like a liberated woman…

"I am woman, hear me roar…"

"It's not a musical!" shouted the King. But the more he covered his ears, the louder the orchestra got.

ARMS

In the absence of guns

June 2000
The American Spectator

CELEBRITY news from the United Kingdom:

In April, Germaine Greer, the Australian feminist and author of *The Female Eunuch*, was leaving her house in East Anglia, when a young woman accosted her, forced her back inside, tied her up, smashed her glasses, and then set about demolishing her ornaments with a poker.

A couple of weeks before that, the 85-year old mother of Phil Collins, the well-known rock star, was punched in the ribs, the back, and the head on a West London street, before her companion was robbed. "That's what you have to expect these days," she said, philosophically.

Anthea Turner, the host of Britain's top-rated National Lottery TV show, went to see the West End revival of *Grease* with a friend. They were spotted at the theatre by a young man who followed them out and, while their car was stuck in traffic, forced his way in and wrenched a diamond-encrusted Rolex off the friend's wrist.

A week before that, the 94-year old mother of Ridley Scott, the director of *Alien* and other Hollywood hits, was beaten and robbed by two men who broke into her home and threatened to kill her.

Former Bond girl Britt Ekland had her jewelry torn from her arms outside a shop in Chelsea; Formula One Grand Prix racing tycoon and Tony Blair confidante Bernie Ecclestone was punched and kicked by his assailants as they stole his wife's ring; network TV chief Michael Green was slashed in the face by thugs outside his Mayfair home; gourmet chef to the stars Anton Mosimann was punched in the head outside his house in Kensington...

Rita Simmonds isn't a celebrity but, fortunately, she happened to be living next door to one when a gang broke into her home in upscale Cumberland Terrace, a private road near Regents Park. Tom Cruise heard her screams and bounded to the rescue, chasing off the attackers for 300 yards, though failing to prevent them from reaching their getaway car and escaping with two jewelry items worth around $140,000.

It's just as well Tom failed to catch up with the gang. Otherwise, the ensuing altercation might have resulted in the diminutive star being prosecuted for assault. In Britain, criminals, police, and magistrates are united in regarding any resistance by the victim as bad form. The most they'll tolerate is "proportionate response" - and, as these thugs had been beating up a defenseless woman and posed no threat to Tom Cruise, the Metropolitan Police would have regarded Tom's actions as highly objectionable. "Proportionate response" from the beleaguered British property owner's point of view, is a bit like a courtly duel where the rules are set by one side: "Ah," says the victim of a late-night break-in, "I see you have brought a blunt instrument. Forgive me for unsheathing my bread knife. My mistake, old boy. Would you mind giving me a sporting chance to retrieve my cricket bat from under the bed before clubbing me to a pulp, there's a good chap?"

No wonder, even as they're being pounded senseless, many British crime victims are worrying about potential liability. A few months ago, Shirley Best, owner of the Rolander Fashion boutique whose clients include the daughter of the Princess Royal, was ironing some garments when two youths broke in. They pressed the hot iron into her side and stole her watch, leaving her badly burnt. "I was frightened to defend myself," said Miss Best. "I thought if I did anything I would be arrested."

And who can blame her? Shortly before the attack, she'd been reading about Tony Martin, a Norfolk farmer whose home had been broken into and who had responded by shooting and killing the teenage burglar. He was charged with murder. In April, he was found guilty and sentenced to life imprisonment - for defending himself against a career criminal in an area where the police are far away and reluctant to have their sleep disturbed. In the British Commonwealth, the approach to policing is summed up by the motto of Her Majesty's most glamorous constabulary: The Mounties always get their man - ie, leave it to us. But these days in the British police, when they can't get their man, they'll get you instead: Frankly, that's a lot easier, as poor Mr Martin discovered.

Norfolk is a remote rural corner of England. It ought to be as peaceful and crime-free as my remote rural corner of New England. But it isn't. Old impressions die hard: Americans still think of Britain as a low-crime country. Conversely, the British think of America as a high-crime country. But neither impression is true. The overall crime rate in England and Wales is 60 per cent higher than that in the United States. True, in America you're more likely to be shot to death. On the other hand, in England you're more likely to be strangled to death. But in both cases, the statistical likelihood of being murdered at all is remote, especially if you steer clear of the drug trade. When it comes to anything else, though - burglary, auto theft, armed robbery, violent assault, rape - the crime rate reaches deep into British society in ways most Americans would find virtually inconceivable.

I cite those celebrity assaults not because celebrities are more prone to wind up as crime victims than anyone else, but only because the measure of a civilized society is how easily you can insulate yourself from its snarling underclass. In America, if you can make it out of some of the loonier cities, it's a piece of cake, relatively speaking. In Britain, if even a rock star or TV supremo can't insulate himself, nobody can. In any society, criminals prey on the weak and vulnerable. It's the peculiar genius of government policy to have ensured that in British society everyone is weak and vulnerable - from Norfolk farmers to Tom Cruise's neighbor.

And that's where America is headed if those million marching moms make any headway in Washington: Less guns = more crime. And more vulnerability. And a million more moms being burgled, and assaulted, and raped. I like hunting, but if that were the only thing at stake with guns, I guess I could learn to live without it. But I'm opposed to gun control because I don't see why my neighbors in New Hampshire should have to live the way, say, my sister-in-law does - in a comfortable manor house in a prosperous part of rural England, lying awake at night listening to yobbo gangs drive up, park their vans, and test her doors and windows before figuring out that the little old lady down the lane's a softer touch.

Between the introduction of pistol permits in 1903 and the banning of handguns after the Dunblane massacre in 1996, Britain has had a century of incremental gun control – "sensible measures that all reasonable people can agree on." And what's the result? Even when you factor in America's nutcake jurisdictions with the crackhead mayors, the overall crime rate in England and Wales is higher than in all 50 states, even though over there they have more policemen per capita than in the US, on vastly higher rates of pay installing more video surveillance cameras than anywhere else in the Western world. Robbery, sex crimes, and violence against the person are higher in England and Wales; property crime is twice as high; vehicle theft is higher still; the British are 2.3 times more likely than Americans to be assaulted, and three times more likely to be violently assaulted. Between 1973 and 1992, burglary rates in the US fell by half. In Britain, not even the Home Office's disreputable reporting methods (if a burglar steals from 15 different apartments in one building, it counts as a single crime) can conceal the remorseless rise: Britons are now more than twice as likely as Americans to be mugged; two-thirds will have their property broken into at some time in their lives. Even more revealing is the divergent character between UK and US property crime: In America, just over ten per cent of all burglaries are "hot burglaries" - committed while the owners are present; in Britain, it's over half. Because of insurance-required alarm systems, the average thief increasingly concludes that it's easier to break in while you're on the premises. Your home-security system may conceivably make your home more safe, but it makes you less so.

Conversely, up here in the New Hampshire second Congressional district, there are few laser security systems and lots of guns. Our murder rate is much lower than Britain's and our property crime is virtually insignificant. Anyone want to make a connection? Villains are expert calculators of risk, and the likelihood of walking away uninjured with an $80 TV is too remote. In New Hampshire, a citizen's right to defend himself deters crime; in Britain, the state-inflicted impotence of the homeowner actively encourages it. Just as becoming a drug baron is a rational career move in Colombia, so too is becoming a violent burglar in the United Kingdom. The chances that the state will seriously impede your progress are insignificant.

Now I'm Canadian, so, as you might expect, the Second Amendment doesn't mean much to me. I think it's more basic than that. Privately owned firearms symbolize the essential difference between your great republic and the countries you left behind. In the US, power resides with "we, the people" and is leased ever more sparingly up through town, county, state, and federal government. In Britain and Canada, power resides with the Crown and is graciously devolved down in limited doses. To a North Country Yankee it's self-evident that, when a burglar breaks into your home, you should have the right to shoot him - indeed, not just the right, but the responsibility, as a free-born citizen, to uphold the integrity of your property. But in Britain and most other parts of the Western world, the state reserves that right to itself, even though at the time the ne'er-do-well shows up in your bedroom you're on the scene and Constable Plod isn't: He's some miles distant, asleep in his bed, and with his answering machine on referring you to central dispatch God knows where.

These days it's standard to bemoan the "dependency culture" of state welfare, but Britain's law-and-order "dependency culture" is even more enfeebling. What was it the police and courts resented about that Norfolk farmer? That he "took the law into his own hands"? But in a responsible participatory democracy, the law ought to be in our hands. The problem with Britain is that the police force is now one of the most notable surviving examples of a pre-Thatcher, bloated, incompetent, unproductive, over-paid, closed-shop state monopoly. They're about as open to constructive suggestions as the country's Communist mineworkers' union was 20 years ago, and the control-freak tendencies of all British political parties ensure that the country's bloated, expensive county and multi-county forces are inviolable.

The Conservatives' big mistake between 1979 and 1997 was an almost willfully obtuse failure to understand that giving citizens more personal responsibility isn't something that extends just to their income and consumer choices; it also applies to their communities and their policing arrangements. If you have one without the other, you end up with modern Britain: a materially prosperous society in which the sense of frustration and impotence is palpable, and you're forced to live with a level of endless property crime most Americans would regard as unacceptable.

We know Bill Clinton's latest favorite statistic - that 12 "kids" a day die from gun violence - is bunk: Five-sixths of those 11.569 grade-school moppets are aged between 15 and 19, and many of them have had the misfortune to become involved in gangs, convenience-store hold-ups, and drug deals, which, alas, have a tendency to go awry. If more crack deals passed off peacefully, that "child" death rate could be reduced by three-quarters. But away from those dark fringes of society, Americans live lives blessedly untouched by most forms of crime - at least when compared with supposedly more civilized countries like Britain. That's something those million marching moms should consider, if only because in a gun-free America women - and the elderly and gays and all manner of other fashionable victim groups - will be bearing the brunt of a much higher proportion of violent crime than they do today. Ask Phil Collins or Ridley Scott or Germaine Greer.

MUSCLE

Couch power

March 2000
The American Spectator

A WEEK OR TWO back, NBC signed a formal instrument of surrender with the National Association for the Advancement of Colored People. Under the terms of the surrender, NBC has agreed to hire at least one minority writer on any series that runs to a second season. The fellows who created the sitcom may not need any new writers, the humor may very well depend on the particular relationships of the existing writing team, but nonetheless NBC will provide them with a minority writer, paid for not by the production team but directly by the network. So, had this system been in place on Broadway, say, half a century ago, the Theatre Guild would have announced that, after a triumphant first year of Rodgers and Hammerstein's *Oklahoma!*, the writing team for the new season would be Rodgers, Hammerstein, and The Notorious B.I.G. and the title would be changed to *Oklahomie!* to be more reflective of our cultural diversity.

I assumed most people would have the same reaction I did: They'd hoot with derision at this latest feeble stage in America's dismal trickle-down apartheid. I assumed a few might even get angry at the notion of racial classification in the televisual arts. After all, whenever it's suggested that a "transgressive" artist might want to scrape some of the elephant dung off the Virgin Mary in order to avoid giving offense, the cultural community rises up as one in order to denounce this attack on free expression. Deciding the composition of a creative team by racial quota is surely equally an assault on artistic freedom.

But no one laughed. And the only people who got angry were various Hispanic, Asian-American, and Native American groups who felt they'd been left behind when the NAACP's Kweisi Mfume barged ahead and stitched up his deal. I shouldn't worry if I were them. It's probably a sliding scale: For the third season, you'll need a Hispanic; for the fourth, a lesbian; for the fifth, Marv Albert in his peephole basque. What kind of sitcom jokes this will result in is hard to say: "A Protestant, a Jew, an African-American, a Latino gay, a pre-op transsexual and a 380lb wheelchair-bound learning-disabled Native American go into a bar and the guy says, 'Sorry, we stopped serving the writing team for "Leave It To Beaver" in the 14th season.'"

But it's no laughing matter. The nation's commentators are largely silent on the matter, no doubt nervous lest their own pasty complexions fall under scrutiny. I got a call from Kweisi Mfume just the other day, threatening an economic boycott unless this column more vigorously demonstrated its commitment to diversity. "But I've just added my first visible minority to the team," I protested. "A black gangsta rapper who writes all my homophobic gags."

Well, our TV networks have been down this path before. It was NBC who, back in the seventies, referred any joke on homosexuality for vetting by a gay dentist in New Jersey. When this curious fact emerged at a conference on censorship, Ed Weinberger, producer of "The Mary Tyler Moore Show," was impressed: "There really *is* a tooth fairy!" he exclaimed. NBC angrily denied the charge that they were in thrall to a gay Jersey dentist on the grounds that, though still resident in the Garden State, he was now a psychotherapist. Besides, networks have always been craven. After advertisers forced the cancellation of his own show, Nat "King" Cole remarked, "Madison Avenue's afraid of the dark."

So let's not fret for NBC. Today the networks are still afraid of the dark, and you can't really blame them: Getting your teeth pulled by a homophobia-sensitive gay dentist is far less painful than attracting the attention of Kweisi Mfume. When the NAACP called a "diversity hearing" in Hollywood in November, CBS sent along President Les

Moonves, but ABC, NBC, and Fox committed the catastrophic *lèse majesté* of dispatching mere vice presidents. President Mfume took umbrage at this and refused to allow them to address the meeting, keeping them offstage and out of sight for two and a half hours until finally the Executive Pen-Pushers had had enough and walked out. Even then, though, the hapless veeps were anxious not to rile the main man. "Some of the initiatives that Les announced today," said the ABC man, "we were also planning to announce" - if only President Mfume had let them.

You can't blame Kweisi for gleefully kicking the suits around. What's more doubtful is what the NAACP hopes to accomplish through this victory. This issue arose last summer when the country's oldest black civil rights group attacked ABC, CBS, and NBC for keeping African-American actors out of their new sitcoms. Network TV, said Mr Mfume, was "a virtual whitewash." He was tired of seeing honky husbands on all-white sitcoms open the front door and go, "Hi, honey! I'm home! Say, what happened to our amusing black next-door neighbor?" "Oh, don't be so silly, darling. You know he was written out last season."

So, aside from complaining about "ethnic purification," Mr Mfume threatened a lawsuit. He wanted the courts to force the Federal Communications Commission to establish "affirmative action" for minorities on sitcoms - the same kind of quota system the NAACP champions so vigorously in other areas of employment. Thus, if an African-American and a Caucasian are both equally qualified to play the part of Lucille Ball's visiting brother from Milwaukee, then the part should go to the African-American actor.

The NAACP also threatened a black economic boycott of these white shows' advertisers. Blacks watch much more TV than white folks. In African-American homes, the television set is on for an average of 70.4 hours a week. That's some average, and Mr Mfume takes a certain pride in the statistic, since it represents, according to him, a lot of "economic muscle" they could bring to bear on the big advertisers. I doubt it. Mr Mfume has been in the racial grievance business so long he seems incapable of looking at the world through anything other than the wrong end of the telescope. The only economic muscle a guy who watches ten hours of TV a day has is with his Domino's delivery boy. He may see more SUV ads than the average SUV owner, but the chances of him rousing himself from the couch and joining their number are pretty remote. In fact, it's unlikely he has any muscle, economic or otherwise: The amount of time spent in front of the box may or may not be unconnected with the disproportionate predisposition of black Americans to obesity, diabetes, etc. But it's certainly not doing anything for the old brain muscle: A recent study in an upscale suburb of Cleveland, puzzled by the way African-American students were underperforming at school, found that middle-class black kids watched twice as much TV as middle-class white kids.

So you'd think, rather than bragging about this statistic, the head of a National Association that exists, after all, for the Advancement of Colored People would be aghast. He'd tell his members to go read a book, take a walk, confirm whitey's stereotypes and become a crack dealer - at least you'll get out of the house occasionally. But instead Mr Mfume thinks the networks should make their shows more appealing to blacks, so that presumably blacks will watch even more TV and maybe smash that coveted 100-hours-a-week barrier. Then he can start railing against executives for deliberately foisting their debilitating video opiate on the vulnerable African-American community.

And so far it's all proceeding to plan. Despite the utter economic irrelevance of Mr Mfume's membership, the big white advertisers were getting nervous, so the networks hastily decided to institute the Mfume quota system. In the days following his complaint, ABC added minorities to five shows, NBC to six. The official line was that they'd been planning to do this all along: The NAACP had just been misled by the sitcoms' whites-only pilots; the nets had always intended to introduce a black best friend in Episode Three. It was all just an unfortunate misunderstanding.

Blacks' favorite TV shows are apparently those of the World Wrestling Federation and Jerry Springer, with his even stagier fights. W E B DuBois and the NAACP's other founders would be mystified by the idea that targeting this electronic narcotic even more precisely at blacks is any way to Advance Colored People. Around the time Kweisi Mfume started his great campaign, Sergei Stepashin, then Boris Yeltsin's Prime Minister Of The Week, was breezing through Washington on a goodwill trip. To be honest, the ruthless ex-Interior Minister doesn't really breeze anywhere - just ask the Chechens - but some spin doctor or other had advised him that, in the debased form of popular democracy practiced in the United States, politicians are expected to show their lighter side.

So he told a joke. A bit of an old one, actually - dating from the good old days of the Cold War, when it used to go something like this:

Soviet guy shows visiting American the latest small Ukrainian car. The American asks how much it costs. The Russian, reluctant to say it's too expensive for most Soviet workers, replies: "And what about your lynchings?"

Not the best gag in the world, but, addressing the National Press Club in Washington, the Russian Prime Minister didn't improve matters with his ponderous re-telling – "It's a small car, very convenient for medium-class people, for instance…"

By the time Comrade Stepashin got around to the punch line, it had evolved into:

"Well, in the States, you kill Negroes."

Washington's official Federal News Service managed to lose a little more in translation than usual and bowdlerized its transcript of the occasion into:

"Well, in the States, you kill beetles."

Negroes or beetles, it left the Press Club cold, and the Prime Minister's foray into stand-up died.

Watching the NAACP/network dispute, Stepashin might have modified his punchline: In the States, there's more than one way to kill Negroes. You could, for example, create a powerful class of race commissars and shakedown artists determined to maintain black people as a permanent victim class.

Wherever he is now, W E B DuBois must surely marvel at a world where the latest battle in the struggle for civil rights is the need to boost the number of hours blacks spend in front of the television set. But then, if blacks switched the TV off now and then, they might get around to pondering whether their interests are really served by so-called community leaders like Kweisi Mfume.

Maggie, Maggie, Maggie! Out, out, out! Really.

November 26th 1990
The Independent

"AND THE DAY'S other news…" began Jon Snow, quite early on in Channel 4 News. By Friday, "other news", which had taken most of the week off, was back in business. The story continued, but it had lost its balls on Thursday. As the tumult abated, you couldn't blame the BBC for blurring the distinction between fact and fiction by surrounding the weekend bulletins with trailers for "House Of Cards", a new drama series scheduled with what Norman Tebbit probably regards as suspicious timeliness. "Nothing lasts forever," reflected Ian Richardson, as he eyed pityingly a photograph of Mrs Thatcher, turned the frame downwards on the desk and then smirked into the camera: a scene the news boys would have killed for and which caught the spirit of the week better than anything else.

Television has explicit procedures to be followed in the event of the death of the Sovereign, but there are no guidelines at all for a coup d'état. As foreign correspondents in banana republics always say, what was most striking was how quickly normality was resumed and the new regime took over. On Thursday, there were four portraits behind the "Newsnight" table: Thatcher, Major, Hurd, Heseltine. By Friday, there were only three. Perhaps somewhere the vocal arranger of the Socialist Workers' Party is even now rehearsing the lads in "Dougie, Dougie, Dougie, out, out, out!", but somehow I doubt it. Worse than *Hamlet* without the Prince, this is "Crossroads" after Noele Gordon, and likely to be received in much the same way by the viewing public.

The lingering impression, though, is still that somehow television missed the story. Like nervous voyeurs, they had fantasised for months about seeing Maggie get screwed by the men in grey suits, only to fail, on the big day, to rise to the occasion. The events of the past week were unprecedented, but, accustomed to presenting everything except Royal occasions as novelty anyway, television news has difficulty grasping the concept of precedent. The constitutional propriety of what was being done rated a minute and a half with Lord St John of Fawsley, during which the supposedly combative Jeremy Paxman made it pretty obvious he had no interest in this area. Channel 4 News even preferred to send its camera crew round to BBC Radio to film Brian Redhead interviewing Kenneth Baker, and, later in the week, to film BBC TV setting up to film Jonathan Dimbleby interviewing Michael Heseltine.

This is perhaps as it should be. After all, in the last week, it's television that's made the running - as Mrs Thatcher demonstrated by her impromptu appearance on the British Embassy steps in Paris, sneaking up behind the BBC's man in the middle of his live stand-up to emphasise that she was still in control of events. If only. Commentators spoke of Mr Heseltine having the "Big Mo", but, in reality, it was a momentum spurred principally by the needs of television, by the requirement to move the story on a stage in each major news strand. The broadcasting of Parliament, breakfast TV, half-hour lunchtime and teatime bulletins, Channel 4 News, "Question Time", "Newsnight": none of these existed at the time Mrs Thatcher became leader. The Prime Minister, the great deregulator, may well reflect ruefully on Harold Wilson's wish for just one TV and radio station, preferably in Downing Street.

In such circumstances, it would have been impossible for the Heseltine bandwagon to ricochet across the schedules like a sports car taking the hairpin bends on the Grande Corniche. It couldn't happen in America, either: there, network news is confined to two chunks per day. Even worse, the proliferating outlets have led to proliferating opinion polls which the craven Tory back benches react to as twitchily as if they were strapped in ejector seats wired up to the Mori and Gallup clipboards.

So the crisis accelerated. Was it only a week ago that John Major was mysteriously "indisposed" and Douglas Hurd was issuing paradoxically straightforward yet opaque utterances? As Moira Stuart announced over the Ramboesque jingle of BBC News, "He's again given Mrs Thatcher his full backing" - presumably because the full backing he'd given her the previous day hadn't been sufficiently well upholstered to prevent it being picked apart by Jeremy Paxman and co for 40 minutes of studio discussion.

"The World This Week" opted for a different approach - the overseas reaction - and talked to Congressman Henry Hyde, who referred throughout to "Mr Hazelteen" - which makes him sound like something soothing you rub in your back, as opposed to plunging in the Prime Minister's.

"Frost On Sunday", like every other programme this week, showed its reporter standing outside No 10. Unlike the others, though, this one had actually been inside - to help her mother pack. "I told her," said Carol Thatcher, "'Mum, you haven't been in a supermarket for twelve years. Shall I get some groceries for Dulwich?' She said, 'Of course, I've been in a supermarket. I've opened enough of them.'"

Under the circumstances, it was a magnificent performance, adding the strange sight of a television journalist reporting on her mother's political downfall to what had already been a week of extraordinary images.

Even programmes apparently unconnected with the great issue of the day seemed strangely pertinent. "Schofield's Europe" is a touch proprietorial, you might feel, for someone who is after all only a children's TV presenter. But, in fact, nice young Phillip Schofield offers a surprisingly distinctive perspective. This week, he was in France, so naturally he interviewed Chris Waddle, the English footballer, and Timothy Dalton, the Welsh motion picture actor, who as Phillip revealed lives in the same street as him in London anyway. Exceeding Mrs Thatcher's wildest dreams, Phillip has stumbled on a European Utopia: a Continent entirely devoid of Continentals.

VEIN

The politics of death

June 16th 2000
The Daily Telegraph

THE FIRST time I met George Dubya Bush was in New Hampshire last year. He was giving a speech. "We need to eliminate the death penalty in America," he declared.

I leapt up from the table and sprinted to the payphone to bark "Hold the front page!" at the editor. But Dubya corrected himself. "The death *tax*. Death *tax*." He grinned at us goofily. "I'm in *favour* of the death penalty."

I'll say. They like frying 'em in Texas - blacks, women, gran'mas, even Canadians, good grief. "That's pretty funny," he chuckled afterwards. "Me comin' out against the death penalty." Yes, indeed. The last time the Governor discussed the subject he did an hilarious routine, complete with whining trailer twang, of death-row inmate Karla Faye Tucker pleading with Dubya for mercy ("*Pleeeaaase*, Governor. Spaaaare maaah liiife"). The punchline? He killed her anyway.

But we're in post-primary mode now, and Governor Bush is one of those candidates who's so vicious, so ruthless, so willing to do whatever it takes to win that the other day he decided not to kill someone. Ricky McGinn had eaten his last meal - cheeseburger, double fries, Dr Pepper; they'd posted it on the Texas Department of Criminal Justice website, which faithfully records Death Row's dietary delights; he was 18 minutes away from lethal injection... and then word came that the Governor had ordered a 30-day stay. Dubya says he's only applying the same principles he's used in the more than 130 executions he's signed off on. But this is the only lethal injection he's halted except for the case of Henry Lee Lucas, who claimed falsely to have killed 600 people and on the one murder they actually nailed him on turned out to have been out of state at the time. In the case of Ricky McGinn, nobody outside the immediate family thinks there's a chance in hell he's not guilty, but the Governor has now decided to order a DNA test and slow down the traffic in the Lone Star State's busy lethal injection chamber.

What's going on here? In the quarter-century since capital punishment returned to America, three sitting governors have won Presidential nominations. In 1988, the Democrats plumped for Massachusetts' Michael Dukakis, Governor of one of the Union's wimpiest states, whose image wasn't exactly improved when Al Gore (in the primary) and George Bush Sr (in the general election) drew attention to a fellow called Willie Horton. Mr Horton was a particularly unpleasant convicted murderer but Governor Dukakis had a relaxed policy of letting him out on leave to terrorise the citizens of other states.

Four years later, Arkansas Governor Bill Clinton wasn't going to wind up as just another sissy-boy Dem who's soft on crime. At the height of his troubles with Gennifer Flowers, he suddenly decided he needed to fly back to Little Rock for the execution of Ricky Ray Rector, a black man who'd killed a policeman and then turned the gun on himself, blowing a large portion of his brain away. Come the big day, Rector was so unaware of what was happening that he decided to save the pecan pie from his last meal for "later".

What's happened between Ricky Ray Rector and Ricky McGinn? Well, in the last five years, the murder rate has fallen by 30 per cent and these days even some Republican governors are queasy about capital punishment: Illinois Governor George Ryan has issued a moratorium on executions. But that's the point: on the Nixon-goes-to-China principle, Republicans can afford to ease up on killing people. For Democrats, it's a far more perilous course: recently, New Hampshire's Republican legislature voted to abolish the death penalty. The Democratic Governor, Jeanne Shaheen, vetoed the bill: she didn't think a Democrat in New Hampshire could afford to look soft on the issue.

So if you're on death row in a year when your Governor decides to run for President, the math seems simple enough: if he's a Democrat, your last meal probably got a little nearer; if he's a Republican, hold the pecan pie, for a little while anyway. And here's my prediction: the first guy to get elected on an anti-death penalty platform will be a fiscally conservative Republican outraged by the fact that the up-front prosecution costs in capital cases are over four times greater than in life-imprisonment cases.

WRIST

Bands

July 8th 1991
The Evening Standard

T
O THOSE OF us for whom the first all-German Wimbledon final tugs the emotions as fiercely as the first all-German European Central Bank, the pre-match build-up caused considerable confusion.

Apparently, the great Centre Court showdown was to be between a player called Stick and a player called Shteesh. Or maybe it was Steek and Shtick. Or perhaps Stitch and Shteek.

But then, ten minutes before Stick-off, Boris Becker appeared, looking for all the world as if he too was playing in the final. And so he was, one plucky loner taking on the rest of Germany's Davis Cup team: Stick, Shteesh, Steek, Shtick, Stitch, Shteek, or any combination thereof.

For the expert analysts on "Wimbledon '91" (BBC2), this was a tricky match to call. Desmond Lynam plumped for "Michael Shtick", Barry Davies for "Michael Stick"; Arthur Ashe favoured "Steesh" and the umpire of the semi-final "Shteek". Mark Cox, Shticking his neck out, came down on the side of "Steek". Silently, many of us brooded on the old Teutonic proverb: "A Stitch in time saves. *Nein?*"

"And so to a Royal occasion for this all-German final," pronounced Dan Maskell, opening his commentary with a real mixed-double of a sentence. The camera alighted on Kitty Godfrey, now 95 but still Britain's highest-ranked world player. In honour of this distinguished champion, Boris Becker lobbed a deferential torrent of spit in her direction. Or maybe he was just offering his own pronunciation of his opponent: Shteeurrsh. Whatever the outcome of the match, Becker looked a shoo-in for the silver saliva salver, Wimbledon's ceremonial spittoon, presented to the player who disgorges the most over the sacred turf. As the match began and his first ball of spit hit the playing surface, the BBC tactfully pulled back to a wide shot, although they later got some marvellous close-ups of third set nose-picking.

For any of his compatriots who might be watching, Becker thoughtfully provided his own colourful German-language commentary. For any English-serving Germans who might be watching, John Barrett did part of his commentary as an English-language translation of a German opera libretto. "But different their serves, Mark?" he ventured. But elsewhere his mind, John.

"Opponents who know each other so well," mused Mark, and you listened carefully to Becker's mutterings in case he could supply the definitive pronunciation of the other fellow's name. In among the "*Mein Gotts!*" and "*Dumbkopfs!*", I thought I heard him refer to "Schteesh". "Unbelievable. Well. 15-30," said Dan.

Unimpressed by the ranting Becker, Dan found himself inclined temperamentally towards Shteek and his coolness under vocal bombardment. Becker would have done well to be mindful of the old Prussian proverb, "Shticks and Shterves may break my serves, but words will never hurt me." During the breaks, Becker continued jabbering under his towel, and Mark and John speculated on his inner conflicts. "You're looking at a tormented soul here," said John.

I don't know whether his sole was tormenting him, but his wrist bands certainly were. By my count, he changed wrist bands at least eight times, the discarded models piling up in the area behind the chair like a charity drive for unaccoutred Third World tennis prodigies. All those empty wrist bands in search of a decent backhand to wrap themselves around. The ball boys offered Becker a choice of balls and he took longer than most guys would to pick out an engagement ring.

The great Dan Maskell, a man of few words but the only commentator able to illuminate the choices facing a player in each rally, eschewed the tormented-soul approach and confined himself to a few observations on half-volleys and backhand returns. "Game Stitch," said the umpire. "Stick leads four games to three." Mark Cox barely had time to recall the semi-final between Edberg and Steek before the match took another dramatic turn. "Game Shtick," said the umpire. "This is what saved Steek," observed John Barrett. Unlike the ranting Becker, Steesh averaged one murmur a set. Speak softly, but carry a big Shtick.

What a strange Wimbledon it's been. Perhaps Becker would have done better to make like the world's Number One women's player, Monica Seles, and call in sick with a mysterious ailment a couple of days before. Still no word on what it is. Maybe she's lost her ability to fake orgasm with every serve. "Uuuuuuuuuugh!" "Eeeeaaaaauuggggh!" I seem to remember Harry Carpenter suggesting she'd been working hard to keep her "obtrusive grunt" under control. I hope not. I was working up a treatment for a new movie called *When Harry Met Seles*, with a particularly embarrassing scene in the Wimbledon strawberry tent.

As for the men's champion, he returns today to a hero's welcome in his nation's capital. As someone once said, "Stich bin ein Berliner."

HAND

Asking for it

February 11th 2001
The Sunday Telegraph

IT'S ALL down to Catherine Zeta-Jones and Michael Zeta-Douglas now. The news that Tom Cruise and Nicole Kidman have gone the way of Bruce and Demi, Alec and Kim, Dennis and Meg has shaken Hollywood to its core. To most industry insiders, they seemed as perfectly matched as Rock Hudson and Doris Day. "This is really the end of an era," pronounced a shaken Vicky Mayer, deputy editor of *Now* magazine. "The Eighties and Nineties power couples are no more." If you're prepared to work at it, not every Tinseltown marriage has to end in divorce - just ask OJ - but with the silver screen's golden couple on the road to Splitsville, concerned observers are worried that, matrimony-wise, Hollywood is now no different from Solihull or Basildon.

Officially, Tom and Nicole split because of irreconcilable differences in height. But for years they've been dogged by rumour. Unlike Bruce and Demi, who were dogged by Rumer, their distinctively appellated daughter, and Scout, their other child. Bruce said they'd grown apart, though that may have been a reference to Demi's implants. Alec and Kim hadn't grown apart, but she dropped several hints indicating she'd quite like to. Last year, Alec threatened to quit America if Bush won the election. "I'd probably have to go, too," said Kim, that "probably" being the first sign that the Baldwin marriage was not as solid as it seemed. As the Florida recount proceeded and the country trembled on the brink of the Bush dictatorship, Kim decided to act immediately, unsure whether North Korean community property laws would be as generous as California's.

After all, not every Hollywood break-up is as amicable as that of supermodel Claudia Schiffer and the television magician David Copperfield, who, accompanied only by a drum roll and an assistant in spangled tights, sawed his assets in half.

Meanwhile, each afternoon across Beverly Hills the real victims of the divorce epidemic, the progeny of the stars - little Rumer and Scout and Satchel and Karma and Snood and Taioseach and Saveloy - stand distraught at the schoolyard gates, looking for the comforting figures of their undocumented Hispanic nannies but instead finding only strange glamorous creatures, vaguely familiar from TV talk-shows, claiming to be their mommies and daddies and suddenly anxious to demonstrate their fitness for joint custody. Tragically, few Hollywood couples are prepared to stay together for the sake of the kids, the exception being Jodie Foster and her turkey baster, though they too are now said to sleep separately, one taking the master suite, the other the kitchen drawer.

Even gay sweethearts Anne Heche and Ellen DeGeneres, Hollywood's leading celesbrities, weren't immune to the trend. Lipstick lesbian Anne and sitcom sapphist Ellen were supposed to be heading to Vermont to avail themselves of one of the state's new same-sex marriage licenses and to conceive a child in the honeymoon suite of some leafy colonial inn. But suddenly Anne left the pre-marital home and was found wandering the neighbourhood in a confused and partially clothed state. Then she flew to Toronto. Fellow passengers said she appeared distraught and bewildered - and not just because she was flying Air Canada. With many Hollywood couples, the rumour is that one of them's gay. With this couple, the rumour was that one of them was straight. And, indeed, since the split Anne has been dating men. It seems her perfect gay marriage was only a facade, a desperate attempt to cover up the secret shame of her potentially career-damaging heterosexuality.

Few of us can imagine the pressures of trying to sustain a relationship in the fishbowl of Hollywood. You go months without seeing your spouse, as filming commitments take you to opposite ends of the earth. Then suddenly your agents accidentally sign you up for the same movie, and you're forced to spend several weeks in close proximity - though not too close, as under the terms of the pre-nup you're not allowed to stand next to him lest people spot that he's only four foot ten. As the relationship counsellors always advise, it's important for busy couples to set time aside specifically for romance, for exploring each other's bodies, for rediscovering that first, urgent, animal passion - just the two of you, plus director, cameraman, boom operator, continuity girl, key grip, gaffer and a couple of friends of the associate producer, all enjoying the quiet intimacy of the contractually-obligated sex scene. It worked for Alec Baldwin and Kim Basinger, writhing naked on each other in *The Getaway* (1994), but unfortunately, by the time Alec came to make last year's Thomas the Tank Engine movie, the producers were far less cooperative, insisting that the only romantic relationship in the picture was between Thomas and a female engine, Lady. (Don't worry, Thomas always pulled out on time.)

Cruise and Kidman were luckier when they made Stanley Kubrick's psychosexual masterpiece *Legs Wide Shut*, though critics complained that in the sex scenes they seemed to be just going through the motions joylessly and mechanically. So who says they weren't a normal married couple?

But what can the big-time Hollywood star do to avoid ending up on Lonely Street, staring into an empty glass and shrugging "Here's looking at you, Kidman"?

Well, first of all, try not to conduct the relationship in the glare of the cameras. Get to know her first in private, in discreet encounters with no one else around. It worked for Woody Allen.

Second, if you're a Scientologist, make sure you don't get mixed up with a kook who's hooked on some weirdo cult like Catholicism. Are you spiritually compatible? When you take her to see John Travolta's *Battlefield Earth*, does she giggle at his hairdo?

Third, if you use a body double for sex scenes on film, why not consider using one in real life, too? Then you can concentrate on the big dialogue scenes - making amusing banter as you're both interviewed by Oprah - while leaving the day-to-day sex stuff to some Equity-minimum extra. That way you'll avoid the complaints of the first Mrs Cruise, Mimi Rogers. According to her, Mimi rogers but Tom is far more reluctant, preferring celibacy in order "to maintain the purity of his instrument."

Fourth, make sure you have something in common: Alec Baldwin and Kim Basinger are both social activists. But Alec's cause is Federal arts funding, while Kim prefers trawling the country trying to liberate every lab rat and guinea pig from their appalling conditions. She might have been more compatible with the famous film star who, according to indestructible legend, had to be taken to hospital to have a gerbil removed from his, ah, orifice. A star whose commitment to animal welfare is such that he opens a rodent sanctuary in his own bottom would surely have impressed Kim.

Fifth, whatever you do, don't take out a big advertisement in *The Times*, as Richard Gere and Cindy Crawford did, announcing that: "We are heterosexual and monogamous and take our commitment to each other very seriously. Reports of a divorce are totally false. We remain very married. We both look forward to having a family." Richard had been attracted to Cindy by her mole - a facial mark, not a family pet, I hasten to add - but, within months of this public declaration, their marriage was over due to the inevitable irreconcilable differences: he wanted a display ad, she said the classifieds would do.

But, at the end of this sombre week, what most anxious couples want to know is: is there such a thing as a successful Hollywood relationship? Why, yes. Liberace enjoyed a very fulfilling relationship with a fetching young man, but only after he made the wee fellow have plastic surgery to look like Liberace. That way he could enjoy the pleasure of making love to himself and, as Whitney Houston has noted, "learning to love yourself is the greatest love of all". Especially in Hollywood.

FINGER

Press to flush

October 1997
The American Spectator

ON PAGE 67 of the September 1st issue of *The New Yorker*, you can see a reproduction of one of the most glorious of human achievements: Michelangelo's fresco for the ceiling of the Sistine chapel. Not the whole ceiling, of course, but only its most sublime aspect - the revelation of God, surrounded by His angels, extending His right hand to Adam: "The Creation of Man." Adam is missing from *The New Yorker*'s page 67. In his place, suspended in the heavens, is a pristine white toilet. God's finger, instead of reaching out to man, is now poised to press the flush button. This is, apparently, "THE BOLD LOOK OF KOHLER®."

Bold as it is, Kohler isn't so crass as simply to lift a work of art. The advertisement is, naturally, a work of art in its own right, entitled "Heavenly Power" and credited to "Michelangelo, Painter; Scott Seifert, Photographer" - an involuntary collaboration, at least on the former's part. Michelangelo would appear to have drawn the short straw on this assignment: for his half of the picture, he had to fiddle around with difficult tints of paint, suspended up in the air, working at awkward angles. On the other hand, Scott simply had to take a photograph of a toilet.

As toilet shots go, it seems perfectly fine - though I regret to say I'm not an expert in the field of toilet art; I'm a Sistine man, not a cistern man. Still, Scott would appear to have done full justice to "The KOHLER San Raphael® toilet." It is, we're assured, "an environmentally friendly toilet that sacrifices nothing when it comes to power and performance. Maybe it's the 2" trapway. Maybe it's the sleek one-way design."

Maybe. But either way, I can't help feeling God's interest in the KOHLER San Raphael® toilet would be fairly limited. From a strict theological perspective, I doubt whether He would have any use for a KOHLER toilet, even with a 2" trapway. But to the advertising industry such concerns are irrelevant: God, like Fred Astaire dancing with his Dirt Devil, is just another conscript to the ranks of celebrity endorsers. When the new deities of our secular media culture warn against "religious extremists," it's worth considering the alternative they're offering: a world where no one at a major company, its advertising agency, or a *soi-disant* sophisticated literary publication sees anything odd - or trivializing, or tacky, or degrading - in Michelangelo's God flushing a toilet.

As it happens, I'm not averse to toilet marketing. The week before *The New Yorker* appeared, I took some overseas visitors to the State Fair in Plymouth, New Hampshire. They were very impressed by the chemical toilets, supplied by Blow Bros and bearing the company's long-time slogan, "We're #1 in the #2 Business." That, surely, is a more persuasive claim for a toilet's efficacy than vague assertions that "KOHLER has the touch®." Blow Bros even has a better 800 number than Kohler: instead of 1-800-4-KOHLER, call 1-800-4-A-POTTY. While we were at the fair, the septic truck arrived to pump 'em out. It also had an excellent slogan: "In This Game, a Flush Beats a Full House!"

Advertising, unless you have the Blow Bros account, is a subtle art. True image is organic; it arises effortlessly from the product, as the funny-shaped Coca-Cola bottle with the squiggly writing did. No corporate overview design consultancy boys could come up with that: in fact, most of them would charge Coke several million dollars just to tell 'em, "Well, for starters, we'll change the logo and get rid of that stupid bottle." But, if all advertising were organic, you wouldn't need ad men - guys who sit around all day trying to figure out what's the best way to pitch toilets to the *New Yorker* readership. And, if God feels hard done by Kohler's agency, it's probably little consolation to know that His own ministers' judgment is even worse. In 1994, for the first time ever, the Church of England, its attendance in steep decline, sought professional help. The duly appointed agency came up with the Anglican answer to Joe Camel and Ronald McDonald - a matey vicar called Jim. If you're wondering "Why Jim?" well, it's an acronym: JIM stands for "Jesus In Me." And so, across England's radio dials and TV screens, Jim made his pitch:

> *The money is diabolical.*
> *The hours are ungodly.*
> *It's a miracle anybody does it.*
> *The Church of England clergy: it's a hell of a job.*

Did it work? No. Three years on, the dear old C of E's descent down the KOHLER San Raphael® 2" trapway of history has only accelerated. Instead of lifting our eyes to God, the church chose a demented downward mobility whose blokey pop demotic was almost pitifully unconvincing.

Kohler, by contrasts, takes the opposite approach. There's nothing wrong with a KOHLER San Raphael® toilet - though, as a personal protest, I'd rather cross the road to the Blow Bros portapotty than have one in the house. But Kohleresque advertising offers a paradox: the higher it aims, the lower it sinks. The more it departs from the "Blast yourself free with Ease-O-Lax!" approach in favor of high-style, high-concept, urbane artistry, the more it debases everything it touches. It's hard to discern any redeeming humor or charm in the collaboration between "Michelangelo, Painter" and "Scott Seifert, Photographer": it's about selling a toilet. And instead of the overpowering beauty of the original and the awe it inspires in believers and non-believers alike, all we see is a peculiarly apt metaphor for our adman's culture: Instead of God reaching out to give life to mankind, He is now reaching out to flush us down His celestial can.

THUMBS

Down

November 3rd 2000
The Daily Telegraph

AMONG OTHER pearls of wisdom, Frank Sinatra once told me he couldn't stand singers who, 20 minutes into the act, undid their bow ties and loosened their collar buttons. It was phoney, he said. It was intended to convey a fake sense of exertion, but instead it just made the guy look insecure.

I think of Frank's sage advice every time some mayor in West Virginia or state senator in Minnesota introduces Al Gore. The first thing Al does is take off his jacket and come to the microphone in his shirt sleeves. Doesn't matter whether it's down in Florida and 87 degrees or up in Maine and a balmy 34, he takes his jacket off. This is one of the little choreographed movements he's been taught in order to indicate that he's "my own man". Every vice-president has difficulties emerging from his boss's shadow, but Al's had more difficulties than most, if only because this particular president casts long shadows and, whenever you think you've wriggled out into the sunlight, you're plunged into gloom again.

This week, Bill Clinton's shadow descended once more: he gave an interview to *Esquire*, full of the usual petulant complaints about how he's apologised to the country for a "personal mistake", so how come the Republicans haven't apologised to the country for putting him through impeachment? Bill is not, as campaign advisers like to say, "on message".

But more revealing - in every sense - than the interview, was the cover photo: a Monica's-eye view of Bill sitting down, with a strangely satisfied smirk on his face and legs wide open, the end of his tie pointing to his zipper like a "Select Desired Item Here" arrow on a vending machine. Come and get it, baby. He's the President and we are all his interns.

Meanwhile, Al Gore is on the cover of *Rolling Stone*. In Washington, several disapproving Beltway insiders have remarked on the pronounced bulge in the Vice-President's pants. On closer inspection, you notice that Mr Gore's thumbs are hooked in the pockets and his hands seem to be pulling the trousers tighter as if to exaggerate the protuberance.

As Tony Blankley put it in *The Washington Times*: "Is this part of some competitive psychodrama Mr Gore and his boss are engaged in?" I don't know about that but the Vice-President has been in the difficult position of trying to straddle two horses: those who like Clinton despite Monica, and those who like Clinton because of Monica. He's trying to be square but hip, a monogamous swinger.

After the '98 Congressional elections, when the Republicans lost seats, and Newt had to resign, and then Newt's successor had to resign, gleeful Democrats dared conservatives to make the 2000 campaign another referendum on Bill Clinton. Wisely, Dubya declined to take up the offer but, instead, rank-and-file Democrats are staging their own referendum on Clinton, and it's not good news.

Why does Gore need to get Clinton in to campaign in California and do commercials on black radio stations? Because for certain parts of the Democrat base Gore is no Clinton.

On the other hand, why is Gore about to lose West Virginia, Arkansas and his own state of Tennessee? Because white rural Democrats are culturally conservative and the party of feminist apologists for predatory fellatio junkies seems increasingly remote to them.

On the other other hand, why is Ralph Nader running so strongly in Washington, Oregon, Minnesota, Vermont and other liberal states? Because the Left has suddenly realised that, now the oral sex has dried up, Clinton is essentially (as he once described himself) an Eisenhower Republican.

As soon as the supposedly sex-crazed GOP stopped trying to nail Bubba's puffy butt, the internal contradictions of Clintonism began to manifest themselves. Too socially liberal for the south, too fiscally conservative for the north - and, of course, too impossibly glamorous for anyone easily to succeed him.

It's not just that Al Gore had to emerge from the big guy's shadow, but that this particular big guy turns everyone around him into pygmies in the shadow of his *Esquire* crotch. There are no Cap Weinbergers or George Schultzes or Dick Cheneys in the Clinton Administration: the only household name is the President's second greatest enabler, Janet Reno, and the rest are a sorry set of non-entities barely glimpsed by the public except on that one occasion when virtually the entire cabinet gave an ad hoc press conference to assure the American people that there was no truth to this Monica business.

In some strange paradoxical way, Clinton has a knack of sucking the life out of all those people who fall at his knees. Al might have seen it coming. It was, after all, not Mr Clinton but his successor as Governor of Arkansas, Jim Guy Tucker, who wound up going to gaol over Whitewater. On Tuesday night, Mr Clinton's successor as Democratic nominee might like to reflect that, if only by comparison, he's got off lightly.

WRINKLES

Gray dawn

October 23rd 2000
National Review

HI, EVERYBODY. A friend of Bob Dole here. You know it's a little embarrassing to talk about ED. Especially for us conservatives, Republicans, whatever. Every four years many of us on the right find we're coming down with a worse case of it. Yet Electoral Dysfunction is nothing to be ashamed of. Unlike many diseases you get when you're old, this is one you get *from* the old. If you're a Republican candidate, talk to your doctor about it and he'll tell you the chances of getting a vote from any of his elderly patients are about as low as drug prices in Saskatchewan.

Old people vote Democratic: Gore's strongest demographic is 65+. The MTV shots, the Gap clothing are just a front. He should be campaigning in checkered stretch pants and guesting on reruns of Lawrence Welk, which he probably did invent. If we cut off the vote at retirement age, Dubya would be a shoo-in. Instead, after our first black president, Gore's going to be our first geriatric president. At the inauguration, his motorcade drivers should ride low in the seat with only their fishing hats visible.

Don't get me wrong. I like old people. I like old movies. I like old songs. I'll take "Moonlight Becomes You" over "Yo, Bitch! Sit On This" any day. I like old broads. I'd rather date Debbie Reynolds than Cameron Diaz. I like old footwear. How come nobody wears spats any more?

And yet… the phrase "the greatest generation" is beginning to stick in my craw like a bottleful of cheap Medicare Viagra before Swingers' Night at the lodge. The elderly are currently the most malign influence on our politics, the beneficiaries of the most outrageous affirmative-action scam in the country, the practitioners of identity politics so ruthless they make the gays look like the Boy Scouts they'd like to be. Statistical fact: America spends more on its old than on its children. When Dubya says, "Leave no child behind," that's code for: Let's leave these whining old coots behind.

But we can't. American electoral politics is currently one nightmare movie with nothing but Jessica Tandy roles. Miss one roadside granny, there'll be another one along in a minute. First, it was Granny D, the New Hampshire granny, walking across the continent for campaign-finance reform. Then she started picking up aluminum cans along the way. No, hang on, that's Granny Winnie, the Iowa granny, who's forced to pick up discarded cans and claim back the deposit to pay for her prescription drugs. Even then, she found she was reduced to sneaking down to her black Labrador's kennel in the middle of the night and stealing the key to his canine medicine cabinet. Oh, no, wait a minute, that's Granny Aitcheson, the Pennsylvania Avenue granny, who's apparently forced to climb into the Lassie suit, get down on her arthritic limbs, and limp off to the vet - presumably because it gives Al a laugh.

Now I like a heartwarming human-interest vignette as much as the next guy, but I'm grannied out. If Granny D had her way, and campaign-finance reform got passed, the only lovable old-timers picking up cans along the road would be Strom Thurmond and Jesse Helms trying to raise enough to scrape together one lousy radio ad to counter the tide of big-time Clymers in the media. As for Granny Winnie, I was initially sympathetic. You know how many Dr Peppers you have to pick up just to get enough for a bottle of Ibuprofen? So I was happy to do my bit and swing through Des Moines, chucking cases of Bud out the window. But it turns out she *chooses* to pick up cans for a living. Her son's a successful businessman, but, although she let him pay for her new roof and her property taxes, she won't accept money from him for prescription drugs because she feels strongly that *you* should pay for them instead. I'm no trained professional, but, if you'll forgive a touch of the Gail Sheehys, my diagnosis would be that Granny Winnie has some weird aluminum-can fetish hitherto unknown to medical science.

True, there are genuine issues of serious concern here: It's outrageous that many seniors are forced to choose between their weekly prescription and Robert Goulet in dinner theater. But the truth is, Granny Winnie doesn't need money from America's working stiffs: Diane Sawyer, Jane Pauley, and Bryant Gumbel are desperate to get her on the air, and if every American senior did just one high-paying network interview, they could chug down all the prescription drugs they want. CBS already runs a commendable AARP pilot program, allowing many seniors to enjoy a little light, undemanding work in their twilight years – it's called, I believe, "the news division" - and there's no reason why this couldn't be greatly expanded.

Instead, Congress is considering a law to allow Americans to import cheap drugs from other countries, such as Canada. It's true drug prices are lower in Canada. Here's something else that's lower up there, too:

Your salary.

Conversely, here's something that's way, way higher:

Your tax bill.

But the life of a Canuck senior is pretty cushy. Not only are the pills cheaper, but so's pretty much everything else. Your apartment's cheaper, restaurants are cheaper, computer software is cheaper. Indeed, life is cheaper, as you'll discover when you're lying on a gurney in the corridor at the Royal Victoria Hospital in Montreal. Not only are Prozac pills a third cheaper, but you'll need twice as many of them to see you through to spring while you're waiting to hear whether you've been scheduled for cancer treatment. In the US, you'll wait, on average, ten days. In Canada, you'll wait 35 to 45 days, maybe more in

Quebec, though eventually they send you down to Plattsburgh, NY, or Burlington, Vt, so you can die in a foreign hospital, unsurrounded by tiresome loved ones.

But hey, it's the good life you've heard so much about from Senator Jim Jeffords and others, isn't it? So don't let me stop you. Matter of fact, why don't you leave right now? That way, you won't be around on November 7th, Florida will return to the Bush column where it belongs, and the free world won't be consigned to eight years of Al Gore because you cranky ingrates don't feel you're being pandered to enough.

Now it's true that in Quebec average life expectancy is 78, a tad higher than in the US. On the other hand, in Albania average life expectancy is 73. Albania is the unhealthiest country in Europe: a decayed, impoverished, lawless, nasty little basket case of a state, where there's no health system to speak of, 98 per cent of the population are chain smokers, the prescription-drug program involves swimming to Italy, and any Albanian Granny Winnie tramping along the rutted highways is unlikely to find many soda cans because, thanks to the late Enver Hoxha's ban on vehicular traffic, there are no cars to throw them out the windows of. (Incidentally, Hoxha's eco-Stalinist Albania is a reasonable approximation of what Al Gore's America would look like if he decided to implement the full *Earth In The Balance* platform.) Yet Albanian life expectancy is closing in on America's, and even Canada's: These old guys are indestructible. Nonetheless, I understand the urgent need to improve life expectancy in the US. God forbid any of the codgers should kick off halfway through Gore's second term, before they've had a chance to vote for Hillary in 2008.

I'm no conspiracy theorist, but I can't help feeling that the geriatric hammerlock on American democracy is part of some vast left-wing plot. Conservatives often complain about the liberal bias of the media, but it seems to me that, as we nitpick over the comments of TV news anchors, we're missing the most obvious liberal bias of all: the ads. To sit through a network newscast is to be exposed to a barrage of incontinence pads, laxatives, and those funny chairs that carry you up and down stairs. I'd never heard of male itch till I watched Tom Brokaw; I'd never had male itch till I watched Dan Rather. The ads on the news are expressly designed to make younger viewers - ie, more Republican viewers - flee in terror to the wrestling channel, leaving only cranky old bastards to follow current events. You're not telling me this is pure coincidence.

Oh, well. Maybe those Floridians will come to their senses. Maybe come the big day they'll pop a couple of Viagra, set the Barcalounger to cruise control, call the nurse over, and never get to the polls at all. Then America's politicians can forget about devising even more ways to kiss up to them, and get on with more pressing issues like Social Security privatization before the worker-retiree ratio widens to the point where one pimply burger-flipper will be supporting entire gated communities. In the meantime, as someone once said, never trust anyone over 30. And, in that vein, let me recommend an old film from the Swingin' Sixties, *Wild In The Streets*, in which the voting age is lowered to 14, rock star Max Frost is elected president and - acting on the principle that you can't trust them - has everyone over 30 dragged off and stuffed with LSD. A non-prescription drug, admittedly, but I'd like to think Granny Winnie would approve.

TORSO

Duck soup

January 8th 1992
The Evening Standard

BARRY NORMAN returned to Britain's TV screens last night with "Film 92 Special" (BBC1). What's a 92 Special? Sounds like a gun. Whether or not he had one in his pocket, he was certainly pleased to see Michelle Pfeiffer.

He started coolly enough, discussing her and Al Pacino's new picture, *Frankie And Johnny*. "You're a movie star. You can't possibly have low self-esteem. So it was a great piece of acting to put that across."

"Thank you," said the former Miss Orange County.

But, displaying his usual incisive interviewing technique, Barry was reluctant to let the matter rest. "So how do you do it? Do you imagine somebody you know?" Being a movie star, Michelle can't possibly know someone with low self-esteem.

Meanwhile, across the room, the esteem was coming out of Barry's ears (apparently, "Film 92" also denotes the room temperature). "There's one scene in the film, which I think is... is actually charming, because of the way it's played, but it could have been quite salacious, when he asks you, you know, to open your robe, so that he could look at your..." Pause. "...naked body."

"Uh-huh," said Michelle in a non-committal way.

Barry seemed to share Al Pacino's opinion on the desirability of viewing Michelle's torso, if anything even more so, but he lacked the open-your-robe direct approach, his strategy being that, if he said she had a beautiful body, perhaps she'd hold it against him.

191

"You and the rest of the world seem to be in some kind of conflict over the question of your appearance," he ventured. "The rest of the world regards you as an extremely beautiful woman, and you think you look like a duck."

"I don't think ducks are unattractive," she ducked.

But in he dived again. "Ducks would have a very hard time if they looked like you, I promise," he said, working up quite an appetite for Duck à l'Orange County.

"I kinda feel like a duck," she replied, a Peking Duck being eyed by a peeking Tom.

"Will you," he pressed, "accept that the rest of the world is probably more correct than you are, that you're less like a duck and more like a very beautiful woman?"

Half-time score: Rest Of The World out for a duck.

By now, my bath was ready so I went off to play with my duck. When I returned, Barry had moved the conversation on to whether Michelle had been handicapped by her looks.

Michelle might more usefully have put the question to Barry, since it can't be every day she meets a man who wears his hair backwards. But I guess there wasn't time what with the many key areas of Michelle's looks which still had to be explored.

"You played the pure beauty in *Dangerous Liaisons*, the sexy singer in *Fabulous Baker Boys*," Barry recalled. But now she was playing a waitress. "What about her looks?" he asked. Her next film is the sequel to *Batman*, in which, for the first time in motion picture history, Catwoman will be played by a duckwoman.

RIB

Tickler

December 11th 1999
The Spectator

L AST YEAR, Senator John McCain stood up to address a Republican fundraising meeting. "You think that was a tasteless joke?" he began, referring to the previous speaker's Viagra gag. "Listen to this one." He then unburdened himself of the following jest:

Why is Chelsea Clinton so ugly?
Because her real father's Janet Reno.

Rimshot!

But the real punchline came in the deafening silence of the American media: like a genteel dowager on the Tube trying to avoid catching the eye of the gibbering derelict with Tourette's syndrome, the press decided not to notice. On the surface, this seems odd. After all, if there's a constant refrain in American political coverage, it's a lament for the loss of civility in public life. When Senator McCain's Republican colleague Dan Burton called the President a "scumbag", the Incivility Police jumped all over him before concluding that his political credibility was damaged beyond repair; when House Majority Leader Dick Armey offered the milder observation that Mr Clinton was a "shameful person" whose credo was "I will do whatever I can get away with", the commentators tut-tutted that the Republican leadership had descended into the gutter but that the faux pas had rebounded on the hapless Armey. Some noted that this was the same fellow who had "accidentally" referred to the gay Massachusetts Democrat Barney Frank as "Barney Fag".

So it's safe to assume that if any other GOP senator had chosen to impugn at a stroke the three women closest (officially) to the President - to say that his daughter's ugly, his wife's a dyke and his Attorney-General's a man in drag – you'd be looking at one dead Republican.

Anyone in public life quickly develops a sense of where the line is - the point you can't go beyond. Around the same time as McCain told his Chelsea gag, it was revealed that the President and Monica had had a good laugh over the so-called "Apple" joke. (Did you hear about the guy who crossed a Jewish-American Princess with an Apple Mac and got a computer that'll never go down?) But even Bill Clinton - the first president to discuss his underwear on MTV – doesn't do blowjob gags in public. Every politician I've spoken to about McCain's joke says the same thing - that, even if they thought it funny, they find it impossible to imagine circumstances in which they'd tell that joke in public, into a microphone, to an audience. John McCain didn't unburden himself at the end of a long night in an obscure Shriners' Lodge out in Hicksville. He did it in mid-evening, in the heart of Washington, at Morton's, a Beltway bastion.

Personally, I find Chelsea rather fetching in a coltish sort of way. But even if you don't I think we can all agree that there are jokes galore to be made about the Clinton Presidency without dragging his poor daughter into it. Yet what did the Incivility Police do? They looked at their shoes. Just about the only American media outlet to bring it up was McCain's local paper, *The Arizona Republic*, which the Senator has declined to speak to ever since his wife, a recovering drug addict, admitted stealing pills from the American Voluntary Medical Team intended for sick Third World children. The paper responded by running a cartoon showing Cindy McCain in a field of starving infants, holding an emaciated child upside-down by the ankle, with the caption "Quit your crying and give me the drugs".

The *Republic*'s readers were not impressed by their Senator's sense of humour. As Ed Childers of Phoenix wrote, "The first thing Senator John McCain did following the telling of a completely derogatory and vile 'joke' at the expense of Chelsea Clinton was to write a letter of apology. If he ever says that about either of my daughters, the first thing he'll have to do is call someone with a big spatula to scrape his ugly ass off the floor."

But to the national press his ugly ass is a thing of beauty and a joy forever. And if that joke didn't persuade John McCain to run for President, it should have. It underlined the essential feature of the McCain campaign - that no "gaffe" can ever undermine the media's investment in him, and that, in such a favourable environment, even being bananas can be a plus.

By the time he was kicking poor little Chelsea around the room, the press had spent months telling the public that Senator McCain was that rare thing, a Republican who cares about America's children, and they weren't about to change the script. "It's like a return to the Kennedy era," one magazine editor said. "He makes a gaffe, and we look the other way." He's the most media-beloved Republican this side of Colin Powell, routinely hailed for his courage, idealism, dignity, integrity, nobility, not to mention his willingness to cook up an anti-tobacco bill even crazier than the one the Democrats were proposing. That's the main reason the press love him: he may be a Republican, but he's shrewd enough to always position himself as the enemy of their enemies, whether Big Tobacco or George W Bush.

His nominal colleagues in the Republican Party find this infuriating, as I discovered at the first McCain campaign stop I attended, at Lebanon Airport in New Hampshire a few months back. It was a breakfast meeting, so Senator McCain, a short, stiff man with white hair, made a brief speech and then took questions.

"Why's he doing all that nervy twitching?" whispered a friend. "Is he on cocaine?"

"You're thinking of George W Bush," I said. The Texas governor was then in the midst of a flurry of drug rumours exacerbated by oddly qualified denials ("I have not done cocaine in the last week and a half").

"Look at his shoulders," hissed my friend.

"That's not coke," I said. "That's dandruff."

"I know that," she said. "But what sort of presidential candidate has dandruff all over him at eight in the morning?"

She had a point. You can forgive the old snowy mantle at the end of a long day's campaigning, but breakfast is different. The official explanation for the dandruff is that McCain was ill-treated by the Vietcong as a prisoner of war and that, 30 years later, he's unable to raise his arms above his shoulders. So he cannot comb his own hair. His aides are obliged to do it for him and, while they're at it, brush down his jacket. As the campaign staff tell it, this is rather a poignant and touching way of defusing the Dandruff Issue. I've come across all kinds of men who bear all sorts of scars of war - from shrapnel to Gulf War Syndrome - but I'd never before met a chap who can honestly say the Vietcong gave him dandruff. I was impressed. It subtly underlined the way McCain is, er, head and shoulders above the other candidates.

But my friend wasn't satisfied. "In that case," she wanted to know, "why didn't they brush down his jacket this morning?" And then all sorts of darker stories begin to emerge - of a volatile, unstable personality who looses torrents of four-letter words at little people for no reason, the sort of fellow whose dandruff you're reluctant to draw his attention to first thing in the morning for fear he explodes.

It turns out that, in an ideologically-riven Congress, John McCain is a truly bipartisan figure: both sides loathe him. There's a persistent rumour that the only reason his fellow Republican senator, Utah's Orrin Hatch, decided to get into the race for President last summer is that he can't stand McCain. Senator McCain concedes that he called another Republican, Iowa's Charles Grassley, a "fuckin' jerk", but says that he and Chuck are now "friends" ("friends" in the context of the US Senate means they have the warm, close, personal relationship of, say, Suha Arafat and the Israeli government). When he was a humble Congressman, *The Atlantic Monthly* reported McCain's altercation in the aisle of the House with Democrat Marty Russo: "Seven-letter profanities escalated to 12-letter ones and then to pushes and shoves." It takes a while to decipher this code but, reconstructing the incident, "seven-letter" is a reference to 'asshole' and "12-letter" to 'motherfucker'. One mayor back in his home state says that he's not happy with the idea of McCain having his finger on the nuclear button.

So on Sunday the Senator released 1,500 pages of medical records proving conclusively that he is not clinically insane - though for my own part I'd like to see what's in the handful of pages that were held back "for personal reasons". But, for the moment, we must accept the word of his doctors that John McCain is not, to use the medical term, stark staring nuts.

Nonetheless, in private many senators agree with that Arizona mayor. They've been hoping that George Dubya Bush's insurmountable lead would obviate the need to go on the record about McCain the motherfucker. But that's no longer the case. In the last month, rumours that John McCain is out of his tree have done wonders for his campaign. A few weeks back he was a respected Vietnam vet with a strong command of foreign policy, and languishing in the single digits. Then he got transformed into a certifiable wacko whose brain had been fried by the gooks back in Hanoi, and suddenly the New Hampshire polls were showing the senator running neck and neck with Bush. McCain had found his issue: it's the insanity, stupid!

Back in Washington, the Arizona Senator's chums in the media had a new line: "McCain's a good fit for New Hampshire," said the pundits. "Like him, they're kind of cranky, prickly, a little, er, unpredictable up there…" In other words, he's nuts and we're nuts; it's a perfect match. "After all, in '96 they voted for Pat Buchanan…" When stories about his marbles finally began to intrude on the McCain-media lovefest, it was widely assumed to be a "whispering campaign" by Bush staffers. Within days, the press were cheering McCain for being canny enough to turn the issue around and make it work for him.

So, throughout New Hampshire, at one campaign stop after another, someone stands up and asks about the rumours that he's explosive and out of control. "Boy," says McCain with mock solemnity, "that really makes me mad." The crowd laughs. "I was just exploding about that earlier this morning." More laughs. "Look, my friends, I get angry sometimes. I get angry when I see Congress wasting billions on weapons systems even the Pentagon doesn't want. I get angry when I see 12,000 of our brave fighting men and women living on food stamps. I get angry when I see the lobbyists and special interests in Washington corrupting our democracy. I get angry when I see gross injustices perpetrated…" Etc.

Actually, there's no evidence that John McCain has ever got angry over any "gross injustice" or matter of public policy. Every incident recounted by Senate colleagues revolves around some piffling perceived slight; mention weapons systems and McCain is perfectly calm, but use the last piece of Senate toilet paper and he calls you a motherfucker. Nonetheless, his carefully constructed response to questions about his temper - a laboured running joke followed by sanctimonious hogwash - is going down a storm with the national press, who marvel at the ease with which he's making the issue play every which way for him. Doubtless there are those who do believe the guy was psychologically damaged by his years as a POW, but who's to say that's a negative? Baby-boomers - not least the middle-class media boys who mostly figured out ways to steer clear of the war - have a weird collective guilt complex about Vietnam. They still think it was a dumb, pointless conflict not worth getting killed for, but, precisely because of that, they have an exaggerated respect for military service. Electing a psychologically scarred POW from a war that America lost would validate both his sacrifice and their lack of it in a way that electing, say, Eisenhower wouldn't.

That's why John McCain's military record is working for him in 2000, whereas, in 1996, Bob Dole's record never did. Dole suffered physical injury in a good cause; if McCain suffered mental injury in a stupid cause, that accords far more closely with media boomers' view of warfare, and also broader American notions of victimhood. After all, if Bill Clinton can parlay an alcoholic pa and an abusive gran'ma, or whatever the latest version is, all the way to the White House, who's to say McCain's dysfunctions won't play equally well? It may be Christmas but, as always in America, there ain't no Sanity Clause.

And, having sewn up the boomer guilt-trippers and the dysfunction junkies, McCain is now wrapping up the sophists. This new proof of his sanity, some commentators say, confirms what they suspected all along - that the so-called "whispering campaign" by the Bush crowd was really a cunning whispering campaign by the McCain crowd about a non-existent whispering campaign by the Bush crowd, all designed to make the Senator look like the victim of a dirty smear and emphasise his military record. And, even if the Vietcong did unglue him, McCain's lost more brain cells than Bush ever had in the first place.

After a month of this, one thing's clear: the Vietcong may not have driven McCain bananas, but McCain does a pretty good job of driving everyone else bananas. He's 100 per cent pro-life, pro-gun, anti-gay, anti-Martin Luther King Day, but droning media lefties hail him as a "maverick", "moderate", "liberal" Republican. "He's 1,000 per cent anti-gay," says Barney Frank, who can't quite figure out why pro-gay, pro-abortion *Boston Globe* columnists are in love with him. Conversely, conservatives seem to have decided that he's a closet leftie who's unreliable on the issues. At the Gun Owners of New Hampshire dinner, Senator Orrin Hatch hinted that we couldn't trust McCain to protect our Second Amendment rights. "You'd be shocked at one of the senators who wants to be a leader in this country… Enough said," he muttered darkly. If McCain's poll numbers get any better, Hatch and his Senate colleagues won't stay silent.

But who'll listen to them? The supposedly "tough" "60 Minutes" news veteran Mike Wallace turned in a mushy valentine to McCain and then announced that, if the Senator won the nomination, he'd quit his job and campaign for him. God help us. If McCain wins the nomination, I'll quit mine and campaign for Gore.

BREASTS

Factory models

January 8th 2001
The National Post

IN LONDON, the talk is all of the burning issue of the millennium: breast implants for the kids. Experts are divided as to when you should get your daughter her new rack: For her 14th birthday? 12th? As a prize for winning the kindergarten sack race? Maybe in vitro? Or pre-conception: is it easiest just to buy eggs from big-breasted women over the Internet?

If your seven-year old asked for new hooters for Christmas, I hope you sat down and had a serious talk with her: "Now listen here, young lady," I always say sternly. "An implant isn't just for Christmas, it's for life - well, ten or 15 years, give or take a leak or two."

Fifteen-year old Jenna Franklin is getting implants as a birthday present from her parents. "You've got to have breasts to be successful," she says, citing by way of example Pamela Anderson, that icon of Canadian nationhood whose component parts are all US-made. She may also have in mind Jordan, former Page 3 Girl of *The Sun* in Britain, who two years ago was a natural 36C, and unhappy about it. "I've always felt flat-chested," she said. "Bigger boobs will make me more confident." So she had an enlargement to a 36D, followed a year later by another enlargement to 36DD, and in November to 36F. As she was leaving the hospital, she complained that they still weren't big enough: "I've got to wait 12 weeks," she said, "then I can have them done again." Friends think her lack of self-esteem knockers-wise may be due to dating Michael Greco of the TV soap "EastEnders", whose previous girlfriend was Linsey Dawn McKenzie, the famous 34GG porn star.

FACTORY MODELS

I'm unfamiliar with Miss McKenzie's oeuvre, but I've seen enough celluloid breasts over the years to discern the general trend: When Hedy Lamarr ran naked through the woods in *Ecstasy* (1933), she had the proportions of a normal woman. By the Fifties, the likes of Lana Turner were more extravagantly endowed but still recognizably human - it was not inconceivable one might run into an approximation thereof at the local soda fountain (which is, indeed, how Miss Turner was discovered). But today most cinematic sex scenes involve actresses so cosmetically augmented they've become a race apart. The question now is whether Jenna and the rest of the world are determined, as in *Invasion Of The Booby Snatchers*, to join them.

Recently, I gave a fairly serious talk to a very moneyed crowd in New York. Afterwards, one fellow came up to me and, after some chit-chat about this and that play, he said that what had really impressed him about me was that he'd seen some woman reading a column of mine on a nudist beach. His wife rolled her eyes. "He's sick with the nude beaches," she sighed. "Drags me along to them wherever we are." The conversation had lurched into territory beyond the parameters of my thesis, but I nodded politely anyway.

"We men are such assholes," the guy said, slapping me on the back. "The problem is there are no four breasts alike. I've seen two the same, but never four." As Christie Blatchford, speaking (as she often does) for the nation's manhood, put it, for guys breasts "don't have to be spectacular. They just have to be." In other words, you'd have to work hard to come up with a pair the average Joe wouldn't be interested in. Men like breasts precisely because there's no standard model: when they made yours, they threw away the mould. Unlike stamp collecting, where there's only one 1856 British Guiana one-cent magenta for every three billion lousy Canada Post 46-cent definitives, with breast collecting every item's a 1932 fivepenny brown Australian George V with the overprinted OS showing multiple crowns and small A with type 7 watermark.

But the plastic surgery business is the Model T of maracas, the Burger King of bazongas, the Hooters of, er, hooters, the globalization of globes. Strictly in the interests of research, I went into a strip joint on St Catherine Street in Montreal the other night for the first time since I was 17. I felt like Rip Van Winkle. Every dancer had the same stick-on breasts with identically mispositioned nipples and appeared to have been dunked in a vat of industrial-strength depilator. I left after ten minutes, feeling queasy. Now maybe they're untypical, maybe they all had 'em done on Medicare, maybe they should have driven down to Plattsburgh, but, even if they're merely the J C Penney knock-off, I don't think the Bergdorf Goodman-level implant is much of an improvement. It is, I would argue, incorrect and misleading to apply the word "breasts" to the bullet-proof chest Demi Moore sported in *Striptease*, anymore than it would be to describe a Pierce Arrow as a Morgan horse who's just been slipped a couple of steroids. Mass-market implants for the Jenna Franklins of the world represents ultimately the death of the breast: global Jennacide.

A couple of years back, with companies reeling under class-action lawsuits, it looked as if fake breasts might be sued into oblivion. Instead, they've become more environmentally friendly. Scientists have replaced silicone with soya-oil implants, so the topless waitress at your local sports bar can now double as a salad-dressing dispenser. Maybe this is what passes for progress in our strange new world, but, if so, count me out. The only time I've ever felt the urge to write to Miss Manners was to inquire about the correct etiquette when you find out, somewhat late in the evening, that your singles-bar stunner is packing a couple of Dow Corning's concrete Bombes Surprises and you want to retire discreetly without hurting her feelings or your ribcage. We're told that the only reason women mutilate themselves this way is for men. So the only way we'll reverse the impending extinction of the natural breast is if we guys are prepared to stand up for 'em. Three years ago in London, *The Sun*, which has been running topless women on page three since 1970, polled its male readers, who voted overwhelmingly to ban implants: as the paper now boasts, "You only get the real deal on Page 3!" *The National Post* does not feature nude women on Page 3, but nevertheless I feel strongly that we ought to show solidarity. Our campaigns for an electable right-of-centre party and non-confiscatory taxation have not been as successful as we might have wished, but I would hope the Campaign For Real Breasts could command the support of all Canadians, both left and right (the politics, not the breasts).

Meanwhile, spare a thought for poor French starlet Lolo Ferrari, a regular on the popular "Eurotrash" TV show, who had her breasts enlarged 18 times to a 54G before committing suicide last spring. Implants are not biodegradable, so one day centuries from now someone could open up her coffin and discover nothing but two huge gel sacs, sitting there atop a mound of dust, an indestructible monument to her pitiful obsession. It would be nice to think that anyone stumbling on such a discovery would have no clue what they were. But who knows? Maybe by then they'll be a familiar sight in every casket.

NIPPLES

Cornish pasty

January 26th 1997
The Sunday Telegraph

MARIANNE Wiggins will come to regret that Cornish limpet. Indeed, it may already be staring out from her seashell collection in silent reproach.

A limpet, for non-shellfish fans, is something which clings to the rocks and can't be prised loose - a state to which Miss Wiggins, to judge from her contribution to the *Times* series on "Love In The Nineties", has now been reduced herself. I know many *Sunday Telegraph* readers no longer take *The Times*, and I confess that I myself, to avoid offending fellow commuters, tend to tuck it discreetly behind my copy of *Penthouse*. So I should explain that the limpet in question was slipped by Miss Wiggins into her lover's coat:

> *When I knew he was in the taxi on his way to Heathrow, I rang him on his mobile. "Without looking at it," I said, "reach into your pocket and tell me what I put in there." "Oh, God," came the answer, "it's your nipple."*

There's more, of course: she faxes an outline of her hand to New York, while simultaneously describing what it is she'd like to do with it; she goes into a Soho sex shop and emerges with some sort of all-in-one vicars'n'tarts outfit, artfully combining dog collar with split crotch.

Miss Wiggins is both a distinguished novelist and the former wife of Salman Rushdie, who must be relieved he never took her on the Iranian publicity tour. ("While in Teheran, Mrs Rushdie is hoping to pick up a split-crotch chador.") But it's the limpet that sticks.

Henceforth, should I ever bump into her on Old Compton Street, I'll be thinking, "Is that a Cornish limpet in your blouse pocket or are you just pleased to see me?"

It's not that she's a woman - though it's worth noting that only a woman can get away with writing this kind of thing these days: a woman planting Cornish limpets around town is "empowering" herself and "celebrating her sexual identity", whereas if I were to go around slipping huge courgettes into ladies' pockets, I'd be reviled as a posturing clod. Nor is it that Miss Wiggins is, as *The Times* puts it, "mature" - though, as Ian Fleming observed, "older women are best because they always think they may be doing it for the last time", and a faint whiff of that desperation certainly hangs over her *Times* piece. But the issue isn't women or older women so much as that foolish group of people who allow themselves to be defined by a specific sexual act. Like a Cornish limpet, you'll never shake it off.

David Mellor, supposedly a heavyweight Cabinet minister, will always be the toe-sucker in the Chelsea kit, beaming with pride at the approving roar of the football crowd, no matter that his sexual inventiveness was mostly the concoction of Max Clifford. The former presidential aide Dick Morris will always be a toe-sucker, too, though he once appeared on the cover of *Time* under the headline "The Man Who Has Clinton's Ear". As things transpired, not only did he have Bill Clinton's ear, he also had Sherry Rowlands' toe, suckling on his favourite foot while dressed up as Rin Tin Tin.

Miss Wiggins would no doubt disdain such comparisons: for her, a faxed hand is infinitely more erotic than a sucked toe. But other people's sex acts are inevitably preposterous. "Because he is a foreign correspondent," she writes of her lover, "and I'm a novelist, we share a certain power to describe our worlds." But she doesn't. When it comes to what she calls "the erotic", the high-toned novelist is even crasser than the lads in the office who invent the letters for the sex mags. When writing about sex, Miss Wiggins is impotent - for, if you happen to find, say, a Cornish pasty more erotic than a Cornish limpet, there's nothing Miss Wiggins can say to persuade you otherwise. How odd that a woman who attributes to herself "a certain power" with words should be turned on by a split-crotch bustier - a generic, production-line notion of sexiness cranked out by some Velcro factory in Basingstoke. Like so much in our liberated world, such sex is Mickey Mouse stuff - literally so in the case of those postcards sold in London newsagents of female breasts with rodent whiskers round the nipples. The vocabulary of sex, its public face has been appropriated by the Max Cliffords and Dick Morrises and whiskered nipples. The rest of us should keep it to ourselves. Bungalow Bill Wiggins - Joan Collins' former love - is fine. But "Barnacle Bill" Wiggins and her Cornish limpet? Put it back on the shelf.

CHEST

Pain

November 16th 2000
The National Post

I AM WRITING this from the front line of our non-two-tier health system - my local CLSC in Quebec. I've been here all morning. I've read *The National Post*, I've read *La Presse*, and *Le Devoir*, *The Gazette*, *Le Journal*. There's nothing left except *The Globe And Mail*. So I figure I might as well start writing.

I came in on Tuesday evening for my 6pm appointment. "There are no doctors after 5pm," said the receptionist.

Really? I examined my official CLSC slip that says "*Vous avez rendez-vous avec Dr...*" Well, let's call him, say, Dr Juan Tier.

Anyway, my little form says "*Vous avez rendez-vous avec* Dr Tier, November 14th, 6pm."

"Oh," said the receptionist. "Dr Tier cancelled all his appointments this afternoon. We left you a message."

"No, you didn't," I said. She looked up my file on the computer, and had good news. "Your appointment wasn't cancelled," she said.

Terrific!

But she hadn't finished. "It was at 9.30 this morning, *monsieur*."

I invited the receptionist to examine my appointment form. It was definitely a "6," not an upside down "9." She called a number, and elicited further information. "No, your appointment was definitely at 9.30," she said. "But, in any case, it was cancelled on October 26th. We left you a message."

"No, you didn't," I started to say, but gave up. I could not see the doctor because a) there are no doctors after 5pm; b) my appointment had been cancelled; c) my appointment had been eight-and-a-half hours earlier; d) my appointment had been eight-and-a-half hours earlier but had been cancelled three weeks ago. We may not have a two-tier health system but we do have a four-tier explanation system. So when could I see the doctor?

December 12th. 6.50pm.

"But just a minute, didn't you say there were no doctors here after 5pm?"

No matter. If I didn't want to wait until December 12th, I could come in in the morning and hang around on the off-chance he could spare a couple of minutes to tell me I only had three months to live and unfortunately I'd wasted two of them trying to get an appointment to see him. This is the system all our political parties have pledged not to tamper with.

Ideologically, I'm an agnostic on health care. I often think the most sensible system is the one used by those unfortunates we Canadians are supposed to feel the sorriest for - the 40 million Americans without health insurance. Most of those 40 million are young adults in their first jobs, and, given that health insurance is a racket bearing no actuarial relationship to the likelihood of your getting breast cancer, kidney failure, a stubbed toe, whatever, or to the cost of treatment thereof, those poor vulnerable uninsured are in fact making a very rational decision. True, if they get some rare, chronic disease, they have a problem. But not that big a problem. A couple I know both had a bad run of luck last year and ran up $40,000 of medical bills, but the hospital agreed to be reimbursed at the rate of $20 a week. That's cheaper than insurance, and considerably less than the average Canadian pays for health in his taxes.

But, for the typical 30-year old, not being insured is not merely rational but liberating: You can walk into just about any hospital in the US and get pretty much anything you need. Any other system - whether governed by insurers or the state – will tend to greater cost, fewer choices and more inconvenience. And in the western world no other health structure costs more, offers fewer options and inconveniences you on a more extravagant scale than the Canadian system our politicians are so insanely proud of.

I speak from limited personal experience. Plagued recently with sundry ailments, on the road and requiring treatment in situ, I've had a chance to experience the varying delights of the Canadian, American and British health systems. As a patient, I've rather enjoyed the opportunity to contrast and compare: There's nothing like getting a second or third opinion. The problem in Canada is that it's very difficult getting a first opinion: We have the worst doctor-patient ratio in the industrial world. Two months ago, motoring along the campaign trail south of the border, I was suddenly crippled up in pain, and so pulled off the Interstate and into the nearest hospital. Ten minutes later, I was lying on a gurney and being put through the usual tests. Half an hour later, having diagnosed the reasons for my excruciating pain, the doctor mentioned to me that there was something on the EKG he didn't like the look of and suggested I get another check when I got home. Back in Montreal the following week, I made an appointment to see the doctor here, asked if I could have an EKG, and, after securing his approval, was given an appointment on October 13th.

That's the one I'm still waiting to hear the results of. In the US, the doctor discusses your EKG with you half an hour after it's done. In Canada, you get your EKG on October 13th and the doctor's available to discuss it with you on December 12th at 6.50pm, assuming the no-doctors-after-5pm rule has been temporarily suspended that evening. Might be nothing. Might be fatal. Might be that the stress of not knowing adds a certain creative tension to the column. Might be that I dropped dead last night and this piece has been cobbled together by my assistant from my last garbled words.

Now I know we're not meant to make comparisons with the great republic to our south, because to move to a US health system would, as we know, be shattering to Canada's national identity. So how about somewhere closer, philosophically, to home? The British National Health Service is the largest employer in Europe, its hospitals have a depressing seediness that will be all too recognizable to Canadians, and, along with Ireland, Britain's doctor-patient ratio is the lousiest in western Europe, though not as bad as Canada's. But the NHS costs the taxpayer less, covers more, and, in any case, there's a private system operating alongside. That, incidentally, is what is meant by a "two-tier system": not the heartless American way, but the system prevailing in Britain, France, Switzerland, Australia and everywhere else in the democratic world. In the great health-care debate, our politicians have carelessly enlarged Canada's visceral anti-Americanism into a lavish, expansive anti-worldism.

At one level, this is appropriate. After all, the Canadian system, of which Stockwell Day is so proud, is shared only with North Korea and Cuba. I can't speak for North Korea, which has been spending its money not on bettering the lives of its enslaved subjects but, as readers may recall, on building nuclear missiles capable of targeting Montreal. But Cuba is an instructive comparison: Fidel has a doctor-patient ratio almost twice as high as Canada's. We have achieved a truly remarkable condition: all the coerciveness of the Cuban system, with none of the efficiency.

With hindsight, Joe Clark crossed out the wrong word in that TV debate. Stock had a little sign saying "No Two-Tier Health Care," and Joe wittily scored through the "no" to reveal the Alliance's secret agenda. If only. Joe's vandalizing skills would have been better employed crossing out the "two." That's what we have in Canada now: "No Tier Health Care." We have no tiers left to shed.

Meanwhile, Jean Chrétien has been forced to concede - oh, horrors! - that he has been presiding over the thin end of the two-tier wedge. On his watch, strange "private" MRI clinics have popped up across the Dominion, mostly for third-party clients, such as insurers. But, like the first tiny growths of a malignant cancer, the Liberals have decided these have to be cut out now before they spread.

Hey, that's great news, isn't it? There are more MRI machines in the city of Philadelphia than the whole of Canada. The average wait for an MRI in the US is three days. In Canada, it's six months - but that's with all these ghastly private clinics. Now that we're getting rid of those, I'm confident we can push that waiting time all the way up to the coveted one-year mark. And that's what counts, isn't it? Universal lack of access. Equality of crap.

You know who I figure could use a cranial MRI? M Chrétien. He's been behaving very oddly lately, telling his caucus he wants to campaign on the issue of prostitution in Brazil and offering to lead the House of Commons in a singalong of "God Save The Queen". Is he clinically insane? I'm sure many voters would like to know, but, sadly, he couldn't get an appointment until, oh, November 28th* at the earliest.

And now, if you'll excuse me, I have to drive to Plattsburgh, clutching my chest and howling in agony. Hop in, Jean, I'll give you a lift.

---

* *The day after that year's general election.*

CIRCULATION

Let's do launch

August 7th 1999
The Daily Telegraph

MADONNA WAS there. So was Salman. And Barbra and Gore and Demi and Slobodan and Gwyneth. I refer, of course, to Tina Brown's perfect launch party for her new magazine, *Talk*, held last Tuesday under the Statue of Liberty, surely Tina's only rival as Queen of New York.

As someone who turned down her application to be *The Daily Telegraph*'s Deputy Sub-Editor (Regional Radio Listings) in 1979, I always knew she'd go far. But I never thought, on a balmy summer evening, with the elegant glow of Chinese lanterns twinkling in the breeze, that I'd find myself wedged into a corner sharing a joke with Idi Amin and Ronnie Corbett. Ronnie was going on a bit, as usual, and Idi was bored out of his skull - a distant cousin of the Kabaka of Buganda that he keeps in his purse as a lucky charm.

So I dragged myself away and who should I run into but my old friend Armand Croissant, the world's most sought-after party designer, with his long-time partner, a beautiful ebony-skinned 17-year-old homeless Haitian who mugged him in Central Park last week.

"Fabulous party!" I said.

"The credit is all Tina's," said Armand, modestly. "She's the high priestess of the Zeitgeist."

"Oh, that's very good," I said, writing it down. "Can I use that?"

"I'm afraid not," he said. "She's copyrighted it."

Armand has an unerring touch for matching each party with the perfect concept, so I was interested to know how he'd come up with such a breathtakingly original theme for Tina's. "I always take my cue from the client," he explained. "So, when Tina told me her magazine was called *Talk*, I immediately thought, 'Hmm. *Talk*. What

212

can I do with that?' And I instantly had this vision of bringing all these people together, clustering them in small groups and getting them to talk."

"Brilliant!" I said. "It's so fresh, so different!" I looked over Armand's shoulder and spied Javier Solana discussing economic reconstruction in the Balkans with Hollywood hunk Matt Dillon. Suddenly, Matt faltered, the conversation trembled on the brink of drying up, and, as if on cue, one of Armand's security team held up a prompt card over Javier's head.

"Um, what's your favourite group?" read Matt.

"I quite like Jefferson Airplane," said Javier, "but, when all's said and done, you can't beat early Floyd."

It's that kind of attention to detail that distinguishes an Armand Croissant party. Alas, for Armand himself, the gathering was not going well. "What is she wearing?" he screamed, pointing across the floor. "Get that dowdy creature out of here! She'll ruin everything!"

"Er, that's the Statue of Liberty," Henry Kissinger pointed out.

"I don't care!" shrieked Armand. "No one's wearing grey this season!"

He was practically hysterical, so I offered to take him home. As we made our way through the crush and past the statue itself, he caught sight of Emma Lazarus' famous words inscribed on the base: "Give me your tired, your poor, your huddled masses yearning to breathe free, the wretched refuse of your teeming shore..."

"My God, what a concept!" he said. "*That's* what we should have done."

Safely back on Manhattan, Armand seemed to cheer up and suggested we look in on another party of his, for the launch of the US edition of *Wad*, "The British Men's Magazine For Blokes Who Don't Give A Toss". High atop Rockefeller Center, the glittering Rainbow Room had been completely redone as a traditional English pub, The Pig And Condom. No sooner had we stepped out of the express elevator than we were greeted by hundreds of pools of simulated vomit, filled with thousands of individual hand-diced carrots. As we entered, an animatronic shaven-headed figure staggered up, grunted, "Oy, mate! You lookin' at my bird?" in a robotic voice and nutted me between the eyes. "*Bravissimo,*

Armand!" I said, as he helped me up from the parquet. "But this must have cost a fortune!"

"Don't even ask!" he sighed. Inside, the famous revolving dance floor had been transformed into a giant rotating pickled Fabergé egg, underneath which the Mormon Tabernacle Choir was singing "'Ere we go! 'Ere we go! 'Ere we go!"

"Exquisite!" said Armand. But already it was time to move on, for he'd promised to look in on yet another of his gala events, the launch of *Launch*, "The New Magazine For Editors Who'd Rather Launch Than Edit". Armand's concept was to stage the event on a rotting launch on the Lower East Side, slowly sinking into the fetid waters as more and more celebrities crowded on to it. "These days," he confided, "it's more important for the party to make a splash than the magazine."

At the end of the wharf, I spotted hot new British editor Sheila Brusque holding up the debut issue - a single sheet of paper, badly photocopied. "*Launch* totally reinvents the concept of magazines for the information age," she yelled over the noise of a police patrol boat. "It's the magazine that says you're hot, you're now, you're in New York. Do you really want to be spending a lot of time going into the office and editing a magazine? Or would you rather just mingle at the launch party with Kevin Bacon?"

But by now Sheila had spotted me. "So what are you doing these days, Mike?" she asked.

"Mark, actually," I said. I explained that I'd just been appointed deputy editor of *Loser*, "The Magazine For People Who Can't Get Into Launch Parties". "As it happens, it's our launch tonight," I added.

"Really?" said Sheila. "I don't recall hearing anything about it."

I didn't think the *Loser* bash would be their scene, but Armand insisted on coming along. It took a while before we could find a taxi prepared to take us there - to a banqueting suite just off the New Jersey Turnpike which had been entirely rebuilt to resemble a Virgin Railways economy-class waiting area (or, as Virgin calls it, Supergold Ultra-Standard Non-Executive Purple Zone). There was no food, no drink, nowhere to sit. "Look here," said a man ahead of us to a Customer Service Representative. "I asked for a Coke 40 minutes ago."

"As soon as we have further information on the beverage situation," she replied coldly, "it will be posted on the electronic indicators."

"They're not working," he protested. But she'd already disappeared into the vast throng.

"This is superb!" squealed Armand, kissing me on the lips. "*Loser* have excelled themselves! This" - he waved his arm airily over the despondent crowd – "this is a party!"

And he was right: ahead of me I could see Nikki Sprong of Chelmsford with her boyfriend Derek, his mate Jason from West Bromwich with his Auntie Beryl and a group from her OAPs' community centre, Ted the security guard with his pitbull Ron…

"*Everybody* is here," said Armand, sick with envy. "*Literally!* Why didn't I think of that?"

ARTERY

Stent

November 24th 2000
The Daily Telegraph

YESTERDAY was Thanksgiving, the day when citizens in 49 states and 64 of Florida's 67 counties gave thanks to God that they had not - as yet - attracted the attentions of Al Gore's lawyers. Meanwhile, the Vice-President's determination to sue till he wins is taking its toll on his opponents. On Wednesday, Dick Cheney suffered a mild heart attack and had a stent inserted in his artery. I've no idea what a "stent" is but it makes a change from "chad". I hope he doesn't wind up with a hanging stent.

In other medical news, his opposite number, Joe Lieberman, seems to be coping well with the emergency integrity by-pass he had to undergo before joining the Gore team: the boys sent him out all over the TV shows this week, defending the decision of Democrat counties to throw out as many as two-thirds of all military ballots, and it didn't seem to be causing him any pain or discomfort in the slightest. Although he may never recover sufficiently from the Gore campaign to lead a "full" or "normal" life, he should be able to make some sort of useful contribution to society - say, as an OJ appeals lawyer.

Al Gore spent the holiday surrounded by his loved ones – that's to say, his ingenious legal team, who tried to keep him up to speed on the status of their various lawsuits.

On Tuesday, for example, they were very pleased with the Florida Supreme Court ruling. But, by Wednesday, they were back in court suing Dade County for, er, complying with it.

Dade's Democrat-controlled canvassing board had more than 600,000 ballots to recount by hand and figured there was no way it could do it by the new Sunday deadline, especially with the biggest holiday of the year in the middle, so it voted to shove the manual recount and certify the (if you can follow this) original recount before the harassed counters all wound up in the bed next to Dick Cheney having dimpled chads removed from their arteries.

As someone "devoted" to "public service", Gore seemed bewildered to discover there were government workers whose idea of a happy Thanksgiving didn't involve bobbing for Democratic votes among the fallen chads. So the Vice-President returned to the Florida Supreme Court to sue Dade into resuming its recount.

Stick to your guns, Dade! Stick to 'em like pregnant chads to a butterfly ballot! Why bust a gut for a bunch of shysters who won't even be grateful long enough to wangle you a couple of invites to the inaugural ball?

If Gore had just named the number of votes he needed, maybe they'd have opened up the stationery cupboard, and dimpled 'em there and then. But, after two weeks, the small signs of increasing grumpiness from low-level Democratic county officials are very encouraging.

If only it were replicated higher up. A few months back, Barbra Streisand listed the three crucial issues of Campaign 2000: the Supreme Court, the Supreme Court, the Supreme Court. What few of us realised was that she was actually talking about the Florida Supreme Court. We should have seen it coming. In recent years the Supreme Court of New Hampshire has effectively overturned the state's entire tax regime while the Supreme Court of Vermont has discovered in its 18th century constitution a right to "gay marriage". In such a world, why shouldn't Florida's Supreme Court shoot for the big one and install in the White House the guy who lost the election?

Earlier this year I wrote in *The Spectator* that increasingly "the judiciary now fulfils the same role as the military in banana republics: you have the right to cast your vote, but the real, entrenched power in the land will see to it that the legislature will only ever be a figleaf".

I thought I was indulging in a bit of rhetorical overkill, but, on reflection, I was underestimating it. In two Florida counties - Palm Beach and Broward - you no longer have the right to cast your vote, only the right to have your vote – or lack thereof - "interpreted" by Vote Interpretationists who are solemnly sworn to divine your true "intention".

Al may be the Western world's lousiest election campaigner, but he has been a brilliant post-election campaigner. Who else could have got a state court to overthrow its own election law on no grounds whatsoever? No evidence of voter fraud, mechanical failure, nothing but Gore's own raw hunger for "chads". If nothing else, the desperate man who launched this insane coup-by-suit has demonstrated how vulnerable the levers of the state are. Even if he himself is cast into oblivion, the monstrous precedents he has established will loom over this republic for years.

LUNGS

War (non-smoking section)

June 3rd 1999
The National Post

WAR IS HELL. Just ask the Bosnians. In 1995, Linda McCartney donated 22 tonnes of dehydrated veggiburger mix to the strife-torn former Yugoslav republic only to discover, after it had arrived, that it contained twice as much fat content as advertised. Lady McCartney was aghast. But she quickly swung into action, organizing a massive relief effort to get the food recalled. The dangers of excessive cholesterol intake in a war zone are well known. God forbid you should keel over from a massive heart attack before Slobo's Serb irregulars have finished mutilating you.

Now the same humanitarian concern has manifested itself at CFB Trenton and the other Canadian bases welcoming Kosovar refugees. Although 98 per cent of the Kosovars, like most Eastern Europeans, are smokers, the Red Cross has declined to provide them with cigarettes. Since the tales of uptight, nervy, twitchy refugees prowling the perimeter fence in search of a sympathetic passer-by who could nip down to Loblaw's for a pouch of Balkan Sobranie first began to seep out, the Red Cross has been at pains to deny that it's subjecting its charges to cruel and unjust punishment. True, they won't supply the Kosovars with cigarettes themselves, and under Canadian law the tobacco companies are in the unusual position of being forbidden to give away a perfectly legal product. But, if Joe Schmoe wants to toss the odd pack of Craven A's over the wire, the Red Cross is prepared to look the other way. Craven, eh?

In a letter to *The Toronto Sun*, the editor of *Tobacco News Online*, Stan Shatenstein, pointed out that what's happening in Balkan villages and what's happening in Balkan lungs aren't so very different. "The tragedy of Kosovo is near unspeakable, but so is that of tobacco. Little bombs explode, one lung at a time, one coronary artery at a time, in one family at a time," said Stan. "The Kosovars are here seeking refuge from quick, ghastly means of annihilation. The Red Cross should not be the agency of a slow, horrid death for the scarred survivors who have joined us here in Canada." What a chump that Slobo is: If only he'd bombarded Kosovo with discount cartons of du Maurier Lights, he could have carried on a slo-mo delayed-action ethnic cleansing, and the West would never have noticed.

At this point, I should declare that I'm not one of those conservatives for whom being militantly anti-anti-smoking is a badge of honour. In Washington during the impeachment trial, I turned up at a restaurant with my colleague David Frum and several other distinguished right-wing crazies. "Smoking or non-smoking?" asked the hostess.

"Smoking, of course!" snapped the fellow from *National Review*. "Preferably a table in a narrow low-ceilinged windowless corridor with Edward R Murrow and Humphrey Bogart in the next booth wreathed in something Turkish and untipped." (I quote from memory.) Naturally, none of us actually smoked.

But I must say, now that the Shatenstein Tendency is so dominant in Canadian life, I'm toying with starting up a 60-a-day habit. The Shatensteins seem to have missed the essential difference between death from Slobo and death from smoking: If you're mooching around the barn one morning when the Serbs pull up, rape your wife and daughters and disembowel you, your diminished life expectancy is involuntary; but if, on the other hand, you decide to nip round the back of the barn for a quiet smoke, your diminished life expectancy is the conscious choice of a free individual. And, indeed, an entirely reasonable choice: the lifetime risk of getting lung cancer from smoking is one in ten. By comparison, in major North American cities, the rate of HIV infection among homosexual men is over 50 per cent, but we don't fret about a one-in-two lifetime risk of getting HIV from gayness. Instead, we hand out condoms. A condom has a failure rate of over 15 per cent - for pregnancy, that is; for sexually transmitted agents, the rate is over 30 per cent. That's not like smoking; that's not a distant deferred risk down the road for a guy with a 20-condom-a-day habit for five decades. That's the risk on the condom you're wearing now. Yet subjecting yourself to that risk is regarded as being "responsible."

Why are smokers the only fellows who can't be trusted to be "responsible"? Well, there's always the risk to others of "second-hand smoke" - a scientifically unproven myth whose promotion by government makes us look every bit as irrational and superstitious as those backward Balkan types. If there were anything to it, the whole of Eastern Ontario would have keeled over by now from second-hand smoke drifting across the border from nicotine-shrouded Quebec.

In fact, the statistics tell their own story: In fastidious smoke-free Ontario, average life expectancy is 79; in Quebec, where we puff away to our heart's content, it's 78. To gamble 12 months for the pleasures of being able to enjoy a relaxing smoke at the end of a night out is a rational human choice. Incidentally, in Albania and Macedonia, where many of those Kosovars who manage to escape CFB Trenton will wind up, average life expectancy is 73. Albania is the unhealthiest country in Europe: a decayed, impoverished, lawless, nasty little basket-case of a state, where, in recent years, smoking has increased by 20 per cent. Yet Albanian life expectancy is closing in on Canada's.

Over here, meanwhile, the aptly named seniors' publication *CARP News* recently reported what it called the "astounding" statistic that 40 per cent of our senior citizens are regular drinkers: This dramatic figure, they say, would place an increasing strain on the health system as our hospitals struggle to cope with alcohol-exacerbated rheumatism, heart disease and dementia. At least any Albanian who makes it past 74 won't have to put up with a lecture from the nanny state every time he cracks open a bottle of Baby Duck. The real threat to our nation's health would seem to be from non-alcohol-exacerbated dementia.

A few weeks ago, billboards went up in Los Angeles showing my old friend Sammy Davis Jr on stage in Vegas: "He took our breath away," said the poster. "The tobacco companies took his away."

Oh, phooey. Sam's life wasn't a tragedy. He smoked it up, boozed it up, had chicks-a-go-go, played Broadway, made movies, topped the Hit Parade, headlined at hitherto whites-only supper clubs. For a short one-eyed ugly black guy of his generation, he had a better life than he could ever have expected. I'd rather have had Sam's 64 years than …well, I was going to name some non-smoking vegetarian nonagenarian but I can't think of any.

I wouldn't be so boorish as to put up a billboard attributing Linda McCartney's early death to her gloomy vegetarian diet. Free-born citizens should be entitled to assess their own "lifestyle choices." Sammy Davis understood his weaknesses: As he ruefully reflected in song, "What Kind Of Fool Am I?" As they pad around CFB Trenton gasping for a smoke, I'll bet the Kosovars are wondering that about their strange, proscriptive Canadian liberators: What kind of fools are they?

LIVER

Cheers, Jenna!

June 3rd 2001
The Sunday Telegraph

I RAISE A TOAST to Jenna and Barbara Bush. Or I would, if I could get a drink around here. The teenage Bush babes are all over the papers for attempting to buy margaritas with fake IDs at a Mexican restaurant in Austin, Texas, last Tuesday night. This comes two weeks after Jenna was ordered to undergo alcohol counselling and perform community service for having been found in possession of a bottle of beer. The judge who passed that sentence has now told reporters that it could be revoked and a more serious punishment sought.

According to Katrina vanden Heuvel, the editor of America's leftie dronefest *The Nation*, Jenna Bush has "a problem". "Our DWI President" – that's Driving While Intoxicated – "has set a very, very bad example for his impressionable girls," tuts Margery Eagan at *The Boston Herald*. "The apples have not fallen far from the tree."

Just for the record, the apples weren't driving, weren't intoxicated, and they didn't fall near the tree or anywhere else, although they might have been walking a little unsteadily and mangling three-syllable words. But, then, so does their dad. And, if Jenna Bush has a "problem", then what does Euan Blair, passed out in his own vomit in the heart of our nation's capital, have?

No, the only "problem" that Jenna has is getting a drink. She and her twin, Barbara, are 19, and, in all 50 US states, it's illegal to drink alcohol under the age of 21. Jenna can drive, vote, marry, own a house, join the army, buy firearms, hop a flight to Vermont with a lesbian, get one of the state's new "civil union" licences and spend the night having as much Sapphic sex as she wants. She can do everything an adult can except go into a Tex-Mex restaurant and wash down her incendiary enchiladas with a margarita. She can buy a gun, shoot up the liquor store and steal the beer. But she cannot walk in and purchase any.

So Jenna and Barbara are obliged to have "fake ID". To the average *Telegraph* reader, "fake ID" probably sounds fairly exotic - the sort of thing you see in thrillers, where the guy needs to get out of town in a hurry, meets a furtive-looking fellow in a waterfront bar, hands over $10,000 in small bills, and says he'll need it by Thursday. But, in America, fake ID is now as common as, well, real ID.

In college towns, getting a false driver's licence is as easy as getting a haircut. If you're a manufacturer of small 2x3" cards or you own a photo booth, you'll be able to retire on the swollen fake ID market. And the economic benefits don't stop there. Fake IDs have prompted the development of machines that can detect fake IDs. The shares of one such company, Intelli-Check Inc, went up 20 per cent on the news of Jenna's latest run-in with the law.

These developments are relatively recent. Until 1984, some states had a legal drinking age of 21, some of 18, and some had no restrictions at all. But then a lunatic control freak in the Federal Transportation Department decided that she knew better than anyone the age at which people could drink.

Although she lacked the constitutional authority to legislate in this area, she had some financial muscle. She informed all 50 states that she would take away the Federal Government's highway funding from any jurisdiction that refused to raise the drinking age to 21. South Dakota went all the way to the Supreme Court, but the crazed regulatory megalomaniac won and took her legal team out to celebrate, presumably with Diet Coke.

The maniac's name was Elizabeth Dole, and two years ago she resurfaced, as a Republican presidential candidate. On the stump, the helmet-haired Mrs Dole conceded that she wasn't happy with the legal drinking age of 21 that she'd forced on the nation. No, these days Nurse Ratched thinks it should be 24. Twenty-four! It would make more sense the other way round: instead of starting drinking at 24, you should stop drinking when you're 24, pick yourself up from the pool of vomit, wipe down your jacket, sober up and start going to work.

Come to think of it, for anyone over 24, the opportunities for social drinking in most parts of America are already pretty minimal. In my corner of New Hampshire, they're virtually non-existent. My mother, who's Belgian and partial to a Stella, was here last year and we swung by the local diner for lunch. She asked for a beer. The waitress looked at her like she was a crack whore. No alcohol. If we'd wanted to, we could have driven 50 miles to the nearest "sports bar" and sat in a basement with guys with no teeth.

I would say that my small town of a few hundred souls is fairly typical: one third are "alcoholics", another third are "recovering alcoholics" and the remaining third are divided between abstemious natives who drink sugary soda and abstemious incomers from downcountry who drink herbal teas in ever more implausible flavours (elderberry pepperoni, etc).

The "alcoholics" buy a case of Bud, drive their trucks deep into the woods and drink it alone sitting on a rock, which is about the only place they won't be given disapproving looks.

The "recovering alcoholics" meet at the library once a month, when they put a big sign out on the road saying: "Alcoholics Anonymous Meeting Tonight 7pm". It's not terribly anonymous - everyone can see Earl's truck parked outside - but then these days most "alcoholics" don't want to be anonymous. All kinds of folks now claim to be "recovering alcoholics", even though, in a typical week, they never drank what the average *Telegraph* columnist gets through by 11am. William Hague would be unelectable here. Okay, I know, he's unelectable over there, too. But what I mean is, if a 14-pints-and-proud-of-it guy entered a Presidential primary, he'd get marginally better press than a serial paedophile.

Everyone talks glibly about "the failure of Prohibition" - meaning the years from 1920 to 1933, when the 18th Amendment criminalised booze and got nothing to show for it but organised crime. But, if you look at the broader picture, the Prohibition movement, which began in the early 19th century, has been a stunning success. Americans today drink far less than they did in 1800, when beer was affectionately known as "liquid bread" and every farm made its own hard cider. The products that especially exercised the Prohibitionists - rum, gin and other "hard liquor" (or "spirits", in the more convivial British designation) - are headed for extinction.

American alcohol consumption is lower than almost any other industrialised nation - lower than New Zealand, Britain, Australia; barely half that of Spain, Germany, Ireland and France. On the other hand, America has 164 alcohol-related support groups per million citizens, 20 times the number of support groups as France. America isn't addicted to alcohol, it's addicted to alcohol support groups.

I have lived in both Britain and America and I have no wish to go down the Anglo-Celtic route, where villages that no longer support a store, post office or church have four packed pubs. I don't miss the baying, mooning, urinating and pavement pizzas. But immaturity comes in different guises. In America, adulthood is so deferred that many Americans exist in a state of perpetual childhood, 300lb toddlers waddling down the street sipping super-sized sodas from plastic bottles with giant nipples. It's at least arguable that it's healthier for Jenna and Barbara to have a couple of glasses of wine than the sugary Pepsis and Mountain Dews that the law all but forces them to drink. Excessive late-teen soda intake may well be the reason why so many chipmunk-cheeked, perky-breasted high-school cheerleaders are bloated lardbutts by 22.

As New Hampshirites know, it doesn't have to be like that. Just across the border in Quebec, they have the same relaxed attitude to alcohol that distinguishes the Catholic countries of Continental Europe. You can drink at 18, the bars are open till 3am, and the *danseuses nues* weigh under 250lbs. The jurisdictions that have the least alcoholism are those in which drinking is most socially acceptable and integrated into family life. In Quebec and France, they enjoy drinking.

In England and Ireland, they enjoy getting drunk. In the United States, they enjoy getting drunk on insane stigmatisatory excess.

It's obvious that Jenna Bush is going to be hounded by the press every time she's within a hundred yards of a cocktail olive. So she may as well become a role model, not for victims of alcoholism, but for victims of the Dole terror. According to polls, the majority of 18- to 21-year-olds have broken Dole's Law in the past month. Mrs Dole's discriminatory, targeted mini-Prohibition deserves to be overturned. Jenna Bush doesn't need alcohol counselling or community service. After the past month, she needs a good stiff drink. And, if she's ever in this part of New Hampshire, I'll happily drive her over the border to Magog, Quebec, and buy her one.

BELLY

Distended

August 31st 2002
The Daily Telegraph

The World Summit kicks off in Johannesburg today, aiming to tackle poverty and protect the environment... It will consume a huge amount of resources and create as much pollution in ten days as 500,000 Africans manage in a year. – The Daily Record

Caviar And Call Girls Find Their Place At The Earth Summit – The Times

'M GLAD I MADE the effort to attend the opening gala of the Earth Summit, truly a night to remember. The banqueting suite of Johannesburg's Michelangelo Hotel was packed as Bob Mugabe warmed up the crowd with a few gags: "I don't know about you," he said, "but I'm starving..." Pause. "...*millions of people!*"

What a master of timing! The canned laughter - an authentic recording of happy Ethiopian peasants clutching their bellies and corpsing - filled the room.

After the chorus of native dancers clad only in packing cases and palm leaves, Natalie Cole came on to sing her famous anthem to industrial development, "Unsustainable/That's what you are", and 65,000 of the world's most eligible bureaucrats, NGO executive council members and BBC environmental correspondents crowded the dance floor to glide cheek to cheek under a glitter ball of premium ox dung specially flown in from Bangladesh. It glittered because of the 120,000 flies buzzing around it, their gossamer wings dappling the international activists below in a myriad of enchanting shadows.

And then I saw her. She was wearing a low-cut dress and had the most fabulous pair of melons. "Holy cow!" I said, as she approached my table. "They've gotta be genetically modified!"

"No," she said, sliding into the chair opposite and giving me a good look at them. "They're all natural." She tossed them to Kofi Annan. "They're for his organic juggling routine." I had to laugh. Sabine Arounde is the Belgian delegate to Unescam, the United Nations Expensive Summits & Conferences Agenda Monopolisers and, lemme tellya, when she's in a room the rising temperatures are nothing to do with fossil fuel emissions.

"We met at Durban," I reminded her.

"Oh, yeah," she said. "The conference on world health."

"Racism," I corrected her.

"Whatever," she said. "This one's more my bag. I'm very into S & M."

"Come again?" I said.

"S AND M. Sustainable Alternative Natural Development Mechanisms," said Sabine.

We were interrupted by the waiter, as oleaginous as a tanker spill. "Will sir and madam be having the Beluga caviar, foie gras, lobster and magnum of champagne?"

"Certainly not!" I snapped. "The papers back home are full of stories about how we're all scoffing the caviar and chugging down the bubbly while just a mile down the road the locals are holding the Distended Belly of the Week competition. In compliance with Foreign Office guidelines, I'll just stick with Set Menu B."

"An excellent choice," he said. "Would sir prefer the mako shark soup or the black rhino confit on a bed of Amazonian mahogany leaves?"

"I'll have the rhino," I said, "followed by the lightly poached panda with a goldenseal salad and two green-cheeked parrot's eggs over easy."

"And would sir like to see the wine list?"

"Just bring me a Scotch and humpback whale oil on the rocks."

As Sabine ordered, she looked coolly into my eyes and Natalie Cole's voice wafted across the room to capture the moment:

> Like a cloud of smog that clings to me
> How the thought of you does things to me...

The orchestra pit had been converted into an authentic replica of a Rwandan latrine and, even as Natalie sang the line, it sprang to life in a hundred dancing fountains of E. coli-infected martini.

"There's something heady in the air tonight," I murmured.

"It's the CO2," purred Sabine.

Four hours later, the exhausted UN lovely, her spent body glistening with the heat of passion, lay back on the shards of her shattered headboard. "Wow!" she whimpered, struggling for breath. "Now that's what I call sustainable growth. You are incredible!"

"UN seen nothin' yet, baby," I said. Yet, to my extreme annoyance, who should burst through the door but everybody's favourite *Guardian* columnist. "You know, of course, George Monsanto," said Sabine, hastily pulling the tigerskin bedspread around her.

"Monbiot!" I said. "I thought you were running away from *The Guardian* to join the gaily pealing fields of Gamo Gofa, where the rude peasant existence is so much more fulfilling than life in the west."

"I am," he said. "I'm on my way to Ethiopia right now. But I just wanted to stop in and thank you for coming here, eating the caviar, drinking the champagne, sucking the praline-flavoured centres out of the individually wrapped Belgian chocolates on your king-sized bed, and blowing all the billions of western taxpayers' dollars. Without your sacrifice, those poor industrialised chumps would have even more money to spend on double glazing, making their pathetic lives even more worthless and hollow."

"You're right," said Sabine. "But I don't know how much longer I can sustain this level of sustainable development conferencing."

"Rather you than me," said George. "I can't wait to be just a happy, laughing Ethiopian fieldhand."

"Better hurry up," said Sabine. "Male life expectancy in Ethopia is 42.88 years."

"Abyssinia," I said, giving George a cheery wave.

SPLEEN

Puking twerps

May 2nd 1996
The Times

ABOUT TEN minutes into Steven Berkoff's production of *Salome* at the National, after the cast had entered and begun talking ... very ... slooooowly ... and moving as if they were walking through wet cement, you gradually heard the audience's collective penny drop: oh, God, they're going to do the whole thing in slow motion.

In that sense at least, *Free Association*, the first volume of Berkoff's autobiography, is satisfyingly (in his word) Berkovian. By "free association", he means that it's not chronological, but jumps back and forth through his last half-century. Like *Salome*, it takes twice as long and has that same wet-cement feeling: no matter how you struggle, the same handful of plays *East*, *West*, *West*, *East* keep coming round again and again. So do the same handful of wives: you assume, from the reference to "my present spouse", that he's a serial monogamist, but the book's anti-chronological form gives his domestic arrangements the air of an erratically timetabled roster of concubines.

The anti-toff stuff is heartfelt. An East End Jew with no further education save for a three-month sentence at a youth detention centre, Berkoff believes that British theatre is as class-bound as British society: in his words, we "doffed our forelock to the guv'nor". Somewhere along the way, he seems to have doffed his forelock right out of his hair to emerge as the bullet-headed scourge of the smug middle-class theatre establishment. He despises the Royal Court ("never met such a bunch of whining, bitchy actors") and the complacent brand-name national companies sucking up all the subsidies ("millions for dickheads"), and it's hard not to warm to a man prepared to denounce such sacred cows of the cosy British theatre as Dame Peggy Ashcroft and Alan Bennett.

"Allies", as he calls them, are fewer in number. At the end of *War And Remembrance* (1986), he buys champagne for the crew, but no one wants to drink with him.

"The pioneer always has to suffer," sighs Berkoff, but some of the indignities are especially cruel. The "Best Of The Eighties" newspaper round-ups don't even mention him: "I feel like those artists in the Third Reich who were cut out of the history books," he frets. The "puking twerps" of *Time Out* do mention him, but only on their "Shit List", despite the fact that he had "almost single-handedly revitalised a large section of the British theatre". He concedes he's not perfect: he is "a sucker for an inferiority complex". There's not much evidence of this in the book, though he does say that *Greek*, "the most exciting play in the West End", did have one fault: he foolishly failed to cast himself in it.

Between artistic travails, Berkoff has plenty of time to denounce Britain more generally. He hates Christmas ("evil telly") and video shops ("usually run by some yobbo and his horror missus") and the Edgware Road ("pus from a wound that ran off the West End") and restaurants ("minimum charge... you can't sit here; lunches only... we've stopped serving") and corner pubs, newsagents and off-licences ("the triumvirate of the soulless British life"). Unlike the guilt-ridden grammar-school ninnies (sorry; this class warfare is catching) in our big theatre companies and British broadcasting, Berkoff is not romantically deluded about working-class life.

For all the spleen-venting, Berkoff is honest about everything except anything that matters. There is a lack of humanity in the book, obscured by its experimental form in much the same way as the form of his productions obscures the lack of feeling there, too. The jacket shows Berkoff's face one half in theatrical make-up, one half naked. This book is only the made-up half.

WAIST

A place at the table

March 1998
The American Spectator

AFTER BLACKS (or, if you prefer, persons of color) and gays (persons of orientation) and women (persons of gender), it was inevitable that one day fat people would also call for an end to discrimination. After all, America has far more persons of bulk than blacks or gays, and almost as many as women. In 1981, the Centers for Disease Control classified 25 per cent of Americans as obese; by 1991, it was nearly 33 per cent; today, it's pushing 40 per cent. Indeed, America now weighs more than any other society in history.

True, it's not yet the fattest country on earth. That distinction belongs to Nauru, a small Commonwealth island in the central Pacific composed chiefly of guano deposits - which somehow seems very appropriate. On top of the guano sit the island's hearty trenchermen. Pound for pound, they may outweigh the average American, but America's got 250 million and they can muster a mere 10,000. Stick the entire population of Nauru in a New Jersey mall and no-one would even notice.

Anyway, fat Americans are now demanding, as Al Gore would put it, a place at the table. "At last," roared a fat activist the other week, "fat people are on the march!" She was, needless to say, speaking metaphorically. But fat people are on the drive, wedged behind the steering wheel with a couple of Twinkies for the road and heading straight for the latest fat rights demo, with a quick stop at the Dunkin' Donuts drive-thru.

The organization leading the battle for "respect for the obese" is NAAFA - the National Association to Advance Fat Acceptance –

who've been weighing in with their thoughts on this year's first trial of the century – that of the California woman whose 680lb teenage daughter was found dead in her room, covered in bed sores and surrounded by empty food cartons. Although she lived only seven blocks from her junior high, Christina Corrigan had been too fat to make the trip. "Fat kids slip between the cracks all the time," said NAAFA's executive director Sally Smith.

I'm sympathetic to anybody except victim groups. I live in the depths of the chilly north-east, and let me tell ya, when it's 40 below, you want something in bed with you giving off a little heat, not a cadaverous stick-insect who's just moved up from Massachusetts. The larger ladies in my neck of the North Country have always, in my experience, had a sunnier outlook on life. But their official spokespersons sound like any old pinched and shriveled identity-group whiners. Perhaps it's something in the diet, but NAAFA members seem to have a special knack for the inapt metaphor. "This case was a shot in the arm for the fat activist movement," said Judy Freespirit. Actually, it's very hard to give fat activists a shot in the arm: try finding a vein. But they've been talking like this ever since *The Nutty Professor*, the 1996 film in which Eddie Murphy, with the aid of a ton of special effects, plays a 400lb man who yearns to be trim and sexually desirable like, well, Eddie Murphy. The film was a huge box-office hit, out-grossing every other comedy in America. The only people who didn't like it were NAAFA. So what exactly had so offended them when they saw the movie?

"We haven't seen the movie," said Sally Smith. "The theaters have no seats big enough."

"That's just one of the ways fat people are oppressed in this society," added her colleague, Patti Horton.

When they first heard about the film, NAAFA wrote to Paramount offering to work with the studio by going through the script and replacing the offensive jokes with ones that were more affirming of fat people's self-esteem. Not surprisingly, the sinister, emaciated studio execs sniggering into their curly endives declined the invitation - probably wisely. As I understand it, a traditional, oppressive fat joke would be: "I wouldn't say my wife is fat, but, when she went into The Gap and said, 'What have you got in my size?' they said, 'How about the delivery truck?'" Whereas a non- offensive joke would be: "I wouldn't say my wife is fat, but that's because I'm a sensitive partner who understands her need for self-esteem."

I have a measure of sympathy for NAAFA, not because of *The Nutty Professor* but because of all the other pictures cranked out by Hollywood. If you look at any movie set in the North Country, the Deep South, the Midwest or pretty well anywhere except a few effete coastal enclaves, the folks in the crowd scenes look nothing like the folks in real crowd scenes. But what really riled Sally and Patti was the implication that Murphy's 400lb professor needed to lose weight in order to be sexually desirable. "Fat people are very sexual," said Sally, with just a hint of menace. Her colleague added that fat people were very interested in exploring their sexuality. Everyone's exploring their sexuality these days. It used to be that people explored the Tyrol or the ruins of Carthage, but go there now and you'll have the place to yourself: "Oh, we thought this year we'd spend three weeks exploring our sexuality," friends say. "It's so unspoiled." Madonna explores her sexuality, Snoop Doggy Dogg explores other people's sexuality (no appointment necessary), so why shouldn't large people? I myself have met at least one large person interested in exploring her sexuality, and indeed mine. I was a teenager at a party in Albany, New York, and the fattest girl I'd ever seen came up to me and said, "Have you ever done it with someone really gross?"

That's not an approach Sally and Patti and their members would approve. Theirs might more accurately be summed up by the banner stretched across a school hall in a Midwestern suburb, "We Celebrate Ourselves!" In America in 1998, just being who you are is a considerable achievement: I get a kick out of me! Thus, last year, one of *Ms.* Magazine's Women of the Year was Nomy Lamm, who, according to Anastasia Higginbotham, is "inspiring a new generation of feminists to fight back against fat oppression" through her magazine.

The titles of fat activist publications tend to be in-your-face - like *Fat! So?*, "for people who don't apologize for their size" - but Nomy decided to cut to the chase and call her magazine *I'm So Fucking Beautiful*, because "that's what I want to hear."

Maybe it would be easier if, in the same way that the handicapped are now "differently abled," fat people could find their own circumlocutory euphemism – "differently weighted"? But then you realize: with this crowd, "fat" is a euphemism. Compared to Sally, Murphy's 400lb professor is slinky; compared to Patti, Oprah at her 237lb peak is positively cadaverous.

Still, there seems to be a market for it. Stuck in Montreal during the recent ice storm and looking for ways to keep warm, I noticed a remarkable number of ads for "escorts" boasting that they weighed 280, 300lbs. For example, Nikki, who moved to *la belle province* from upstate New York, has 58-inch hips and makes a very nice living. "Tell me," I said. "Are most of your clients men with emaciated 250lb wives looking for something a little ✓meatier? Or guys with 400lb wives in need of something svelter?"

"Mark, sweetie," she sighed. "You sound like you've been spending too much time with these bony-butt French women they've got up here." She has a point. Canadians, obsessed with differentiating themselves from Americans, claim their national identity lies in such distinctively un-American activities as universal health care. It's an unreliable strategy, at least while Hillary Clinton's grand project is still sitting in a desk drawer somewhere in the White House, for it suggests that, in the event that America moves to a socialized health system, Canada loses its *raison d' être*. A far more useful guide as to which side of the border you're on is the width of the rear ends at the shopping mall. Canadians may drive American cars and watch American TV, but, hampered in part by the dietary constraints of ruthless confiscatory taxation, they do not yet have American butts. Nikki herself concedes that her unique selling point would be far less unique south of the border.

The statistics bear her out. Class Three Obesity - which means, broadly, people who can't squeeze into theater seats for *The Nutty Professor* - has risen some 350 per cent since 1970. Every couple of weeks or so on the local news shows, you'll see a group of volunteer firefighters cutting the wall off the house of some fat guy who's gotten too fat to get out of the door. For example, there was 800lb Michael Hebranko, who had to be forklifted out of his home in Brooklyn. When he returned - this time through the front door - he said, "I've learned a lot about me. I learned about keeping my ego in check." He's also learned to keep his hazardous waist in check - dropping (according to one paper) 250lbs or (according to another) 320lbs and now weighing (according to a third paper) 500lbs or (according to a fourth) 600lbs. He'd once weighed 1,000lbs, then lost 700, then put on 800. Because of the inconsistent figures for his inconsistent figure, I tried to call him up, but I got his answering machine, on which he advised: "Stand up from where you are now and move forward and say I will and I did."

Unfortunately, more and more Americans have difficulty standing up from where they are now and moving forward. A couple of years ago, in Texas, a husband, puzzled that his wife didn't seem to be making much progress with WeightWatchers, followed her to the weekly meeting and discovered that she was instead chowing down at a local steakhouse. Confronted by him in bed that evening, she reached for her gun and shot him dead. She was about to get up and skip town when she remembered that one of her late husband's more useful attributes was that he helped her get out of bed. Unable to manage it on her own, she could only lie there helpless as the police arrived to arrest her.

So I'm with Eddie Murphy on this one. The National Association to Advance Fat Acceptance is really a National Association to Accept Fat Advancement - and fat is advancing through American life quite fast enough as it is. You can talk about "fat genes" and "hormonal problems," but the vast bulk of America's vast bulk is due to eating too much, to Double Whoppers and supersized fries. Indeed, given the fuss government now makes about societal costs incurred through smoking, it's surely only a matter of time before it turns its attention to regulating our diet. Fat people who are too fat to do their jobs are now claiming the right not to be sacked by size-ist employers. They insist that they're entitled to their full salary just for being propped up against the workbench and wheezing for eight hours a day. Even the most *dirigiste* libertarian has to ponder whether, when your population is too fat to work or wage war or do anything much except sprawl on the couch and dial the local Domino's, this isn't a matter in which the state has a legitimate steak - er, stake.

Besides, fat may mean sexy, but it also means poor. Generally speaking, the wealth of any town is inversely proportional to the width of the butts filling up at the gas station. At the top level, according to Dr Michael Hamilton of Duke University, "a typical businessman will earn $1,000 a year less for every one pound he is overweight." You can't be a fat cat if you're fat, cat. That's what NAAFA objects to: it's not good for the social fabric to have an unrepresentative, attenuated elite at the top and a huge heap of lardbutts underneath. One day, warn Sally and Patti, the corpulent masses will rise up and storm the barricades, or at any rate lean on them.

Well, let 'em try. They can put in seats that'll accommodate Sally and Patti - and every movie theater will go belly-up. Already the US economy loses $100 billion from weight-related sickness. We may gloat over the "Asian flu" currently afflicting the Far Eastern economies, but what chance has America in the long run, if Sally and Patti get their way, that it can ever compete with those wiry Filipinos and Koreans? The only really important economic question of our time is whether the European Union can hold off collapsing under the weight of its Franco-German social programs until the United States has collapsed under the weight of its weight. On current trends, 75 per cent of Americans will be obese by 2050. Dr. John Foreyt of the Baylor College of Medicine in Texas has calculated that by the year 2230 every American will be obese. That presumably will be when the last anorexic supermodel decides she's fed up getting stared at when she steps out the limo on Madison Avenue and moves to Paris.

In other words, if ever there were a demographic segment less in need of a lobby group, it's this one. So, when NAAFA gets upset by Eddie Murphy movies, I say to them: lighten up.

GUTS

Tell it like it is: we're awful

August 5th 2000
The Spectator

Philadelphia

"THE REPUBLICAN Convention," I said to my cab driver, "and make it snappy."

Fat chance. He pulled up by a stand offering free mammograms. "This can't be right," I said. But he insisted it was the official entrance. "As part of the Convention," said the lady, "we're offering free mammographic consultations."

"Thanks all the same, but that's not much use to me, is it?" I chuckled.

"You'd be surprised," she said, and explained that male breast cancer was America's silent killer. "We're also doing free prostates," she added.

"As long as I don't get Bob Dole's," I said, pressing on before she could start feeling for lumps under my breast and I had to explain that that was my old, gnarled, conservative hard heart.

I know next to nothing about Philadelphia except that Frank "Big Bambino" Rizzo, police chief and mayor back in the Seventies, ran for election on a pledge to "make Attila the Hun look like a faggot". My kind of guy, even though he was a Democrat.

But this week the Republican Attilas descended on the City of Brotherly Love and did their best to look like faggots. George Dubya Bush rode in, his spectacular mound of corpses left far behind in

Texas. His new sidekick, Dick Cheney, a "widely respected bipartisan moderate" until two weeks ago, is now referred to by Dems as "The Man Who Voted Against Freeing Nelson Mandela". They say this portentously, like a movie trailer: "Dick Cheney *is* The Man Who Voted Against Freeing Nelson Mandela. Parental Guidance Advised." I don't know why freeing Nelson Mandela falls within the jurisdiction of a Wyoming congressman, but maybe he spent the late Eighties moonlighting on the Cape Town Board of Paroles.

But that was then and this is now, and now compassion is all. As Dubya likes to say, "Don't judge me by how many people I fry, judge me by what's in my heart." (I paraphrase.) If Cheney is in the mood to jail anyone, he's keeping it to himself. This year's Convention programme is like a bad edition of "The Black and White Minstrel Show": the stage has mostly been abandoned to vapid white women and black men who seem strangely unconvincing. The speakers are heavy on single moms who believe in local control of education, Hispanic agricultural workers who believe in sensible health-insurance reform, breast-cancer survivors who believe in tax cuts, etc. If Nelson Mandela were to stroll by the Convention today, Cheney would be the first to say, "Hey, if you're the African-American senior who believes in strong missile defence, you're not on till tomorrow night."

Thus, the opening day's Pledge of Allegiance was led by a blind mountain-climber. A senior Republican tells me that this is the first blind mountain-climber ever to address a political convention, though no doubt the Democrats are scouring the land for a blind mountain-climbing lesbian. The *soi-disant* uncompassionate conservatives - the pro-lifers, gun nuts, religious right - are unimpressed at seeing the GOP's big tent turned into base camp. "This blind mountain-climber," said one cynic. "What's he climbed?"

"He's climbing Everest next year," I said.

"Hmm," he said, thoughtfully. "Well, I'll be interested to see if he can make it up to the podium."

In the event, he had a seeing-eye dog, who presumably gets into the old goggles and crampons and accompanies his master everywhere, the first K9 on K2.

And who's to say the Republican Convention isn't more gruelling? Scaling the vast mountains of compassionate verbiage, blinded by the lightness, desperate for a glimpse of an Abominable Snowman or two. But Pat Buchanan and Pat Robertson and all the other incendiary Pats of yore are nowhere to be seen: the only Patness to be found is in Dubya's responses. After watching Bill Clinton steal most of their policies, Republicans are now appropriating Bill Clinton's style. After all, he's spent much of the last eight years performing free mammograms on Kathleen Willey and co, and it seems to have worked out well for him.

It is, though, one of the paradoxes of "compassionate conservatism" that it is incredibly uncompassionate towards conservatives. In fact, for conservatives, as well as for persons of taste and discrimination, this week has been hell. As someone on the podium was droning on about how if we taught all our children to fish no-one would go hungry, I passed Senator Mitch McConnell, who looked like someone was twisting a pineapple up his bottom. Okay, he always looks like someone's twisting a pineapple up his bottom, but, from the pained expression on his face, Monday's was a particularly large genetically modified one.

I even caught sight of Newt, skulking round the back of some New Hampshire delegates as if he was worried I might call security. "It's very different," said Bob Dole. "You don't have Buchanan. And you don't have Newt." And, let's not forget, you don't have Dole, with his disastrous 1996 speech promising to build a bridge back to the early 20th century and his home town in Kansas. Bob Dole loves small-town Kansas so much he lives in an apartment in the Watergate complex: Maybe he got tired of a loser burg where there's nothing but a windswept grain elevator and a lunch counter full of grouchy guys like him nursing their coffee. But, wherever the party is now, it's definitely not in Kansas anymore. The Wizard of Austin has performed his magic and the grim monochrome of Dole's Kansas is a riot of colour.

"So what do you think of it?" I asked an Oklahoma delegate.

"Well, George Dubya Bush is a strong candidate with a positive messa…"

Oh, come on, what do you really think?

"Well, there's some truth to the argument that in recent years we Republicans have perhaps tended to be a bit strident and nega…"

No, no, what do you really think?

He sighed. "I keep telling myself all I have to do is get through till the end of the week."

Poor fellow. Like many rank-and-file Republicans he felt strangely excluded from the inclusiveness. The party that opposes quotas has turned its Convention into the biggest quota program of all. In a convention awash with visible minorities, my friend was part of the Republican Party's non-visible majority: he's not black, he's not gay, he can't speak Spanish, he hasn't triumphed over male breast cancer. "Leave no child behind," says Dubya, incessantly. But leave all these cranky white guys behind: that suits us just fine. That was more or less what happened to William Hague, who turned up here to salute Dubya, as one colossus of global conservatism to another, and to the best of my knowledge never made it past the mammogram. I'm no pollster but I would say that about five per cent of the delegates believe passionately in compassionate conservatism, five per cent are longing for Pat Buchanan to walk in and relaunch his 1992 cultural war for the soul of America, and the remaining 90 per cent would prefer cultural warfare but are willing to string along with this ersatz Clintonian shtick if it works.

Finally, though, someone at this Republican Convention had the guts to stand up, go negative and make a vicious personal attack …on Republicans. In his keynote address, General Colin Powell said enough with the niceness and argued that Republicans were racist hypocrites who'd abandoned the mantle of Lincoln and cared nothing for helping black children while being happy to provide corporate affirmative action for fat cats.

And the Republicans loved it! General Powell had said much the same thing at the '96 convention, but the cranky Doleful GOP of four years ago had been unimpressed and had even had the temerity to venture the occasional boo. This time round, they applauded, they cheered, they said to themselves: "You know, that guy's got us just right. We really are awful, aren't we? And he's so right about how great those racial quotas are! If we'd had racial quotas in the GOP, the party wouldn't be fun of unlikeable guys like us!"

Al Gore and his reprogrammers must be baffled. If the Vice-President had wanted to beat up on Republicans for wanting to leave black kids in the ghetto, his advisers would have said, "Nah, that's too negative. Affirmative action's not popular. The speech'll backfire and you'll get hammered for being mean. But maybe we can sell it to the Republicans."

I may have to revise my opinion of Convention honcho Andrew Card's three-pronged strategy of filling the screen with fashionable identity groups; saying nothing beastly about the Democrats; and being unsparingly critical of your own failings. Presumably the idea is that in two weeks' time in Los Angeles viewers will tune in, see a bunch of blacks and gays attacking the Republicans, and go, "Hey, this is a rerun. What else is on?"

And, when you think about it, the GOP's decision to stage the Democratic Convention two weeks early is a stroke of genius. The Democrats run effective conventions ('92, '96), the Republicans run effective governments (most states and most major cities). Dubya has shrewdly recognised that these are two quite separate skills, in much the same way as parliamentary monarchies distinguish between the dignified and efficient parts of the constitution. The Republican Convention is basically the US equivalent of the Queen Mother's 100th birthday bash, if you can imagine the Queen Mum as a black Latina Queen Single-Mum. And, just like the Royal Family, the Convention is scrupulously non-political. Why, on Tuesday morning

Colin Powell went on TV and said he'd be willing to serve in a Gore Administration.

And I don't doubt it. General Powell is far too grand for anything as grubby and vulgar as politics. Would he make a good Secretary of State? Who knows? I can't really see him facing down Slobodan Milosevic across the negotiating table, can you? Maybe he'll be in the cabinet, or maybe after November we can all forget him for another four years until it's time for the next quadrennial refrain of how the General is America's Last Best Hope. We can certainly forget about John McCain and Elizabeth Dole: You wouldn't want McCain as Defence Secretary 'cause he's nuts and you wouldn't want Mrs Dole as anything because she's a big-government dominatrix whose bright ideas include raising the legal drinking age to 24. If there's one thing I've learned during this prolonged exposure to "compassionate conservatism," it's that the right to drink yourself into a stupor is one of the most precious of liberties.

So make the most of Liddy, General Quota, Mister Maverick, the blind mountain-climber, Snoop Doggy Dogg, the full Broadway cast of *La Cage Aux Folles* and the rest of this unlikely rainbow coalition while you can. Come November, I don't expect the Republicans to have made significant inroads among African-American voters. And, whether or not the General knows it, that's not his job. His job is to make significant inroads among those "moderate" white voters who feel vaguely squeamish about the GOP and need to be reassured that pulling the lever for the Republicans doesn't mean you're Jesse Helms.

And General Powell accomplished that. At the last Convention, when Mrs Dole did her Oprah walkabout, that's all she was appropriating: the walk, the style, the shtick. Four years later, they've decided to go the whole hog: not just the roving microphone, but the sensibility - the party was in denial, but now they're in recovery and demanding respect for dealing with their problem. Of course, on the daytime talk-shows, the transvestites who date their wives' fathers tend to acknowledge their problem and then go out and do it again,

245

and that's what the Republicans are planning, too. When the Hispanic nephew, the gay congressman and the mariachi singers have doffed their sombreros and left the stage, what's left is Dubya, Dick Cheney, missile defence, tax cuts and social-security reform. Talk-show platitudes plus conservative policies: The Sally Jesse Helms Show?

NAVEL

Gazing

March 2nd 1999
The Wall Street Journal

A FEW YEARS back, I used to write for *The Independent*, a "quality" London broadsheet. One night my editors sent me to a restaurant, at which, for reasons I forget, I was to be photographed being serenaded by a singing chef. Unfortunately, no staff photographer was available, so they got hold of a freelance. He turned up with a collapsible ladder, which had somehow de-collapsed and so snagged a handful of dangling earrings and repositioned a couple of toupees as he made his way across the room.

When I asked why he'd brought a ladder to a restaurant, he replied that he did most of his work for the tabloids and, as several Royals ate in this place, he liked to keep his ladder with him in case one turned up, because, especially if it was Di, it always helped to get a couple of feet off the ground so you could get a good shot of her knockers. (That's Fleet Street tab talk.) As he warmed to his theme, the more squeamish of the adjoining diners discreetly signaled the *maître d'* and asked if they could be moved from their prestigious power table to the poky little one with the wonky leg wedged in by the restroom door.

I think we can all agree that the tabloid boys are not one's preferred dining companions. It would be perfectly frightful to be invited to an intimate Washington *soirée chez* Katherine Graham and find that, due to some ghastly misunderstanding in *placement*, Kay had seated you next to the sleazebag with the ladder. But the thing is this: The sleazebags got the story. For years, they told us that the Prince and Princess of Wales' marriage was a fake, that the Prince was cheatin' and the Princess wasn't eatin', and we at the high-toned end dismissed it as regrettable, intrusive, speculative fantasy. We were wrong, the ladder guys were right.

So, for all the ponderous, agonized media navel-gazing over the rampant tabloidization of America, the consumer would be well-advised to consider his own interests. To put it in a nutshell, if he wants to know what really happened, he should check out the sleazebags. Thus, when the Princess died and the bigshot yawneroos on the US networks and the broadsheets lapsed reflexively into another round of ponderous, agonized navel-gazing about whether the media had driven her to her death, the Fleet Street tabloids were getting on with the dull job of finding out who actually did drive her to her death: one drunken chauffeur and his cocky Eurotrash playboy boss.

Likewise, OJ Simpson. While ABC, CBS, NBC, *The New York Times* et al couldn't wait to upgrade a grubby little double-homicide into another episode in their ongoing saga of ponderous, agonized navel-gazing, *The National Enquirer* was sticking doggedly to pertinent matters like the whereabouts of the missing knife, the bloody glove - hard specifics that require journalists getting their shoes a little muddy.

We shouldn't be surprised then that President Clinton's impeachment trial went nowhere. Between the Senate and the mainstream media, this was the Super Bowl of self-importance. Like Gene Kelly in *Singin' In The Rain*, both bodies' watchword is: "Dignity, always dignity." When it comes to deans of dignity, solons of solemnity and grampas of gravitas, for every Senator Robert Byrd, I'll raise you a Daniel Schorr. Both parties were too busy gazing lovingly into their own mirrors to concern themselves with the rising tide of presidential DNA lapping about their ankles.

Even now, on the discussion shows, nine out of ten exercises in ponderous, agonized navel-gazing invite as one of the participants Steven Brill, editor-in-chief of *Brill's Content* and one of the great comic figures of the age. Mr Brill is best-known for excoriating ABC News reporter Jackie Judd for her gross irresponsibility in breaking the utterly unfounded story of the stained dress when there was absolutely no evidence such a dress even existed. When it emerged that the dress did in fact exist and was on its way to the FBI crime lab, Mr Brill revised his criticism of Ms Judd. He was no longer accusing her of getting the story wrong, but of getting the story right too soon and for the wrong reasons. The responsible journalistic thing to do would be to triple-check your sources a couple of more times and sit on the story until, oh, sometime midway through Al Gore's second term. In a healthy media culture, instead of swanking about on "Nightline," Mr Brill would be foaming at the mouth and climbing the walls in his padded cell.

Enough with the dignity; never mind *Singin' In The Rain*; it's time to start ringin' in the sane: America's mainstream media need more tabloid values, not less. This isn't really anything to do with public interest - though, in the case of the Prince of Wales, when the heir to the thrones of Britain, Canada, Belize, Papua New Guinea and sundry other realms is effectively perpetrating a massive fraud on his future subjects, that's certainly a matter of public interest. So it is in a republic in which the spouse of the President is accorded an office and staff, when the First Marriage is a sleazy travesty and the President, to maintain the sham, embarks upon a thuggish campaign of character assassination. I'm not talking about the independent counsel's investigation, or the legal requirements for impeachment, or even the core definition of news - something that the parties involved don't want you to know about. It's more basic than that. A fellow who isn't interested when the President of the United States is rumored to be at best a serial trouser-dropper and at worst a rapist has a fatal lack of curiosity, and curiosity ought surely to be journalism's minimum entry qualification.

It's the dopiness of the non-tabloid media that's offensive. I've lost count of the number of times I've woken up, switched on National Public Radio's morning news and heard at the top of the bulletin: "The President travels today to [insert state here] to unveil his proposals on [insert issue here]." That's not news, that's a press release some duty editor has entered in the diary.

In both Britain and America, politics and celebrity are remorselessly converging. But there's one fundamental difference: The British media treat the Prime Minister the same way *The Sun* treats Michael Jackson or George Michael. If there's a rent boy or transsexual hooker or even an unpaid parking ticket, they want to get hold of it. By contrast, the US media treat the White House the same way, say, Diane Sawyer treats Barbra Streisand - with the appropriate sympathetic deference.

Vanity Fair, for example, recently ran David Kamp's lengthy article on the grim trajectory of the "tabloid decade" - from Joey Buttafuoco to OJ to Monica. Nice piece. Wasn't the cover story, though. Nor could it ever have been, no matter how good it was. Mr Kamp's other piece that month - an "exclusive first look" at the new *Star Wars* movie - made the cover. In the complex negotiations to secure access to A-list Hollywood, publications willingly trade cover status, an opportunity to screen the story beforehand and an advance agreement on the parameters of any interview.

Which would you say was more dominant in the American media? The trend toward tabloid sleaze? Or toward monarchical deference? Look at last week's newsmagazines: instead of running with "Is Our President a Rapist?," they put the alleged rapist's wife on the cover in regal pose and cooed over the unstoppable momentum of her mooted Senate campaign.

The standard defense of the respectable, house-trained news outlets is that all the polls show the public is tired of the media's obsession with scandal and wants it to get back to the real issues. Like what? Slobodan Milosevic and his cheery entrail-gouging, breast-severing, genital-skewering Serb butchers? Wake me when they reach Brooklyn. The *soi-disant* Monica "feeding frenzy" wasn't substituting for careful, considered assessments of foreign policy: We all know that the odds of a US network broadcasting a major prime-time investigation into North Korea's nuclear program are less than zip.

The Canadian papers recently reported that the North Koreans have now got missiles targeted on Quebec, on the grounds that nuking

Montreal would demonstrate to the Americans that they're serious but, unlike blowing up Cleveland, would not invite instant retaliation. When the mushroom cloud drifts across the St Lawrence and Vermonters start giving birth to two-headed babies, Dan Rather will turn up in a decontamination suit and ask if it's something to do with La Nina. But until then, Kim Jong-Il's got no chance of making prime time unless he flies Monica over and persuades her to drop to the Pyongyang broadloom. So let's not pretend that covering the latest travails of President Buttafuoco is distracting the press from its natural high-mindedness. Indeed, the respectable media have achieved a weird distinction: They're boring without being serious.

The main reason is that, unlike the tabloid reptiles, the uptown boys seem to have an aversion to work. It's so much easier to convert a story to symbolism and ponder endlessly whether the fall of a black role model is a tragedy for the dream of American racial harmony. No, it wasn't. It was a tragedy for one specific woman and a friend who happened to be with her at the wrong time. Months before anyone had ever heard of Johnnie Cochran, the respectable press had played the race card. Even Mr Cochran's comparison of the Los Angeles Police Department to Adolf Hitler doesn't seem so ridiculous when you consider that one early *New York Times* piece on OJ managed to drag in Norman Mailer, Jean Genet, Dostoevsky, Milton, Shakespeare and Sophocles. Or as the deathless prose of *Newsweek* put it: "Was this another case of power/money/fame's wretched song of impenetrability?"

We should have known from Nicole Brown Simpson's quick consignment to the sidelines how easily Paula Jones and Kathleen Willey and Juanita Broaddrick would be drowned by the upscale media's own wretched song of impenetrability. I'll take the tabloids any day.

THE PELVIS

All shook up

August 15th 2002
The National Post

TOMORROW will mark exactly a quarter-century since a 911 call from 3765 Elvis Presley Boulevard sent the paramedics scrambling for the ambulance. And we're still no nearer knowing the truth.

Did Elvis Aron Presley, appointed by President Nixon as a special narcotics agent, disappear into the FBI Witness Protection Program? Was he murdered by the Mob? Is he living in a rented room above a bar in Flint, Michigan? Is he, as the motion picture *Men In Black* suggests, a space alien who simply "went home"?

Whichever theory you incline to, the Establishment has gone to extraordinary lengths to persuade us that Elvis is, in some sense, "dead." Nine years ago, when the US Postal Service issued what was to become its all-time best-seller, the Elvis stamp, customers angrily pointed out that you're not supposed to be on a stamp unless you're deceased. Only the other day, the actor Nicholas Cage married Lisa-Marie Presley and let it be known that beforehand he'd gone to a medium with whom he contacted the spirit world and received his father-in-law's blessing. How can you talk to someone in the spirit world who isn't there yet?

But, whatever Elvis' current status, it's clear that August 16th 1977 marks a dividing line in his career. After the funeral, his manager, Colonel Parker, reassured Elvis' father: "Nothing changes." Instead, everything did.

Before 1977 Elvis was the first pop star of the rock'n'roll age, the white boy who sang black, etc, etc. His significance in purely

252

musical terms has been most incisively analysed by the dean of rock critics, Greil Marcus:

> *That Elvis did what he did - and we do not know precisely what he did, because "Milkcow Blues Boogie" and "Hound Dog" cannot be figured out, exactly - means that the world became something other than what it would have been had he not done what he did, and that half-circle of a sentence has to be understood at the limit of its ability to mean anything at all.*

Exactly. But, for all his scholarly dissection, Marcus represents old-school Elvisology. By 1977, at the age of 42, Elvis had wrapped up whatever musical contribution he was going to make. Puffy, bloated and over 250lbs - which, believe it or not, was considered large in 1977 - the King was having difficulty squeezing into the rhinestone-studded jumpsuits; he'd forget the words, and giggle through love songs. Not all celebrities age well - or, in this case, middle-age well - but, even when Sinatra or Ella Fitzgerald had an off-night, you could usually rely on the band being competent. In the final decade of his performing career, it's not just that Elvis is covering "My Way", it's that he's covering it with what sounds like a pick-up house band from open-mike night at Cactus Jack's sports bar every Tuesday out on Highway 27, turn right past the abandoned grain elevator.

In the quarter-century since, the albums have continued to sell, but indiscriminately: the early Sun sessions (the stuff Greil Marcus likes), the hymns and gospel songs, "Old Shep", the compilations of Sixties movie numbers like "No Room To Rhumba In A Sports Car", it all sells equally well. The Elvis industry has no interest in quality control.

But it's not about the music any more. Elvis is, in every sense, larger than that. He was always a spiritual man: when he died on the toilet at Graceland he was apparently reading *The Face Of Jesus*, a book about the Turin Shroud. A mere 25 years later, the King himself is generating a brisk trade in sacred relics. Most prized of all is the wart supposedly removed by his doctor when he entered the army in 1958.

The Elvis wart is the Presley fan's equivalent of the Turin Shroud, though, unlike the Shroud, there seem to be several competing warts, many of doubtful provenance: I myself have seen and held what most experts regard as the Elvis wart, personally shown to me by Elvis curator Joni Mabe, who suggested the DNA within it could be used to clone a new King. Less valuable are Elvis toenail clippings, mainly because, from the number in circulation, he'd have to have had the fastest-growing toenails in history. Most are alleged to have been discovered deep in the thick crimson broadloom with which, in 1974, Elvis and his then girlfriend plastered Graceland.

There is a reason why Elvis uniquely inspires this devotion. American celebrity has a basic trajectory: you're born poor, you make it big, you move to a penthouse on Central Park West and a beach house at Malibu, and you acquire a taste for fine dining, expensive risible contemporary art and the other habits of the conventionally rich. Elvis never did. He was born in 1935 in a two-room shack in a dirt-poor section of Tupelo, Mississippi – just about as low as white folks can go. But no matter how rich he got, his tastes never changed. Being rich meant doing all the same things he'd done when poor, only more so: he had banana pudding every night; instead of eating one cheeseburger, he'd eat six; instead of cruising Main Street for a late-night diner, he'd hop on the private jet, burn $16,000 worth of fuel and fly to Denver for a peanut-butter sandwich; instead of a 22-inch TV, he had the planet's biggest set; instead of grumbling that there was nothing on, he'd blow the set apart with his M-16 automatic rifle; instead of shooting beer cans off the tailgate of his pick-up, he'd buy up every available flashbulb in Memphis, toss them in the pool and shoot them out of the water. At Graceland, he took an antebellum colonnaded fieldstone mansion and turned it into the world's largest trailer.

This isn't the decadence of Hollywood and Manhattan, just the regular tastes of poor rural Southern whites on an unlimited budget. No wonder they love him for it: "white trash" - an enduring pejorative even in our sensitivity-trained age - are the most despised social group in the United States, but Elvis never broke faith with them. Like regular Americans, he had no use for abroad, venturing no further than Canada and, strictly for his stint in the army, Germany. Even his fat, much mocked by urban sophisticates, was a sign of solidarity. It was his record producer, Felton Jarvis, who pinpointed precisely his old friend's place in the culture: informed of the King's death, he said sadly: "It's like someone just told me there aren't going to be any more cheeseburgers in the world." In the Fifties, he may have briefly represented rock'n'roll rebellion, America's counter-culture. But he endures as the perfect emblem of America's lunch-counter culture.

Death clarified all this. It was his widow Priscilla who turned Elvis into a brand name. In life he was a most naive superstar: there were no shrewd investments, no offshore funds - just a million bucks sitting in the same kind of no-interest chequing account a minimum-wage waitress would have. His most valuable copyrights had been sold to RCA for a pittance, and the reason he never did any overseas tours turned out to be because Colonel Parker, who claimed to be the son of West Virginia carnie folk, was actually an illegal immigrant from the Netherlands and didn't have a passport.

To secure Lisa-Marie's inheritance, Priscilla set about the belated professionalization of her husband's career. He was posthumously all shook up. She opened Graceland, so that fans could make their pilgrimage to his home and his grave, which now attracts more annual visitors than President Kennedy's. Before Elvis, it was an established legal concept that "the dead have no rights." Priscilla decided to reclaim exclusive rights to her late husband – his image, his identity. The Celebrity Rights Law, passed by Tennessee in 1983 and since taken up by other jurisdictions, effectively extends to Elvis' estate the rights of a living person. Whether you believe he and Osama are working the night shift at the Dubuque Burger King is up to you. But, in the legal sense, Elvis is most definitely alive; it's just that he's changed his name to Graceland Enterprises Inc.

Whether dead or living, whether returned to his sender or at an address unknown, it's hard to know what Elvis himself would make of his transformation. As Greil Marcus would say, he has become something other than he would have been had he not been whatever he was when Greil Marcus was trying to figure it out. His most passionate recording - bitter even - is "Long Black Limousine": the story of a country girl who sets off to make it big in the city, sells her soul and returns home, as promised, in a swanky car - which turns out to be a hearse. After 25 years, Elvis' hearse cruises on unstoppably.

CELL

Empty

July 21st 2001
The Daily Telegraph

Motoring offences are three times more likely than burglaries to end in a prosecution, it was revealed yesterday… Home Office figures showed seven per cent of burglaries resulted in a prosecution, compared to 22 per cent of motoring offences. Labour Minister Charles Clarke admitted in the Commons it was easier for police to prosecute a motorist caught for speeding than it was to pursue a burglar, who had often vanished by the time police arrived. – The Daily Mail

Operation Napkin, begun in Gloucester last month to crack down on diners expressing unsavoury opinions in Indian and Chinese restaurants, used pairs of plainclothes officers eating out on Friday and Saturday nights. – The Guardian

AS THE CLOCK struck six, Hercule Poirot rose to his feet and bowed theatrically. "Mesdames, messieurs," he began, "please arrange yourselves comfortably." I could see the Colonel glance nervously across the library at his lordship, his lordship's much younger second wife, the second wife's penniless brother, the penniless brother's drug-fiend uncle, the drug-fiend uncle's shifty Arab business associate, the shifty Arab's friend the American heiress, the heiress's embittered spinster companion, the spinster companion's Central European armaments dealer, and a few other suspects to make up the numbers.

"So you've solved the case, Poirot!" I cried. "Come on, out with it, old man! Who committed the crime?"

"Ah, *mon cher* Hastings," sighed my old friend. "Always obsessed with - 'ow you say? - 'oodunnit. But you forget, *mon vieux*, that I am now with the South Midlands (North) Police Force and, when it comes to ze crimes, we have ze clearing-up rate of just under four per cent. So, when you say to me 'oodunnit, I can answer only that I have not the faintest idea. I have been back at the station with my feet up on the desk sipping a *tisane* and watching the CCTV monitors in the High Street."

"But then why have you brought us here?"

"Yes, what's the idea, you jumped-up little foreigner?" snapped Colonel Ridgeway.

"*Exactement!*" said Poirot. "You may not know, *mon Colonel*, that I have been doing the plainclothes duty at the Leamington Spa House of Curry, where I order the crème de menthe and stare at my patent-leather shoes and eavesdrop discreetly on neighbouring tables for evidence of racist remarks. On Friday night, you and Lady Windlesham were eating the chicken vindaloo and I 'eard you say you were reading an awfully good Agatha Christie."

"That's right. *Ten Little...*"

"*Sacrebleu*, Colonel, not again! So I use my little grey cells and I think, 'This man, he has the hot temper.' And next day Kevin, who lives with his *maman* on the estate just past the mini-roundabout, comes to see me and he says, 'M Poirot, I was breaking into Ridgeway Manor the other night, just to nick the old git's video, and, as I'm carrying it out across the ha-ha, I heard this voice behind me.'

"According to Kevin, you shouted, 'This is the sixth bloody time you've done this, you tattooed yob', and then, having dialled 999 and received the pleasant voice telling you you're being held in an automated queueing system, followed by an instrumental recording of 'Windmills Of Your Mind', you hurled the mobile phone at Kevin in a rage, striking his nasal piercing and causing him a severe nosebleed that prevented him breaking into Mrs Southwood's house. So I am arresting you for aggravated assault, my dear Colonel."

As Poirot led the Colonel away, his hard-boiled colleague Mike Hammer came in. "Busy night, old chap?" I asked.

"It was one of those evenings when the fog comes down and wraps itself around the world like a five-dollar whore at the end of a slow week," he said. "All I saw was the dame standing there under the sodium light in a dress that was too tight last year. The cold, clammy night glistened on her full, round breasts. 'Help, help,' she whimpered. 'I've just been attacked. They pulled me out of my car and stole my handbag.' Now I knew why I'd noticed her breasts. Her buttons were ripped off, and those babies were coming out to play.

"'Relax, honey,' I told her. 'Call the Violent Assault Hotline during office hours, and we'll send someone over to take a statement early next week.'

"'But they're in the next street, dividing up the cash. If we hurry, we can catch them.'

"I slapped her hard. 'I don't hurry, baby, except when I'm on my way home with a ham and pepperoni and doing 120 in a residential street.'

"She pressed herself against me and the heat of her skin seared my shirt. 'But you're Mike Hammer.' I could feel the rise and fall of her bazoongas against the bruises on my ribcage. I'd been manning the random breathalyser checkpoint, and some punk accountant had opened the door of his Mondeo too quickly. This tomato was better than anything the doc had prescribed. 'You're the hardest-boiled dick on the South Midlands (North) Force,' she purred in my chest hair. 'I hear you killed a couple of guys.'

"'Yeah, but only when I was doing 120 in a residential street. And you should have seen the paperwork afterwards.'"

Mike was interrupted by the return of our Community Liaison Officers, Holmes and Watson. "The entire area has been plagued by break-ins, the village hall is full of angry locals demanding answers, and yet you say you have the solution," said Watson. "What is it, Holmes?"

"The answer is simplicity itself. We shall announce that we are sending an extra patrol car through the village once a fortnight."

"You never cease to astonish me, Holmes! However did you solve this mystery?"

The great detective stretched out in his chair. "Elementary, my dear. What's on?"

"There's a repeat of 'Dixon Of Dock Green' on UK Gold."

"Perfect."

HIP

Square is the new cool

May 17th 2001
The National Post

THE LAST TIME I said something nice about Perry Como it wound up in *The Washington Post.* Joe Queenan was writing about the alleged phenomenon of "conservative cool," having found himself on ABC's "Politically Incorrect" opposite various leggy blond vixens in tight mini-dresses who were hot for missile defence. Evidently rattled by this, Queenan was relieved to discover that there was still one uncool conservative: me. "In the course of a wonderful diatribe arguing that Frank Sinatra is a million times cooler than any contemporary rock star," he wrote, "Steyn went out on a limb and said that Perry Como has made better records in the past 20 years than the Rolling Stones.

"Now, anyone can go out and say that Frank Sinatra has made better records in the past 20 years than the Rolling Stones - in part because it's true - but it seems to me that making the same bold claim for Perry Como expresses something at the very heart of true conservatism. Real conservatives don't like groups called Smashing Pumpkins. Real conservatives don't wear leopard-skin skirts. Real conservatives tell the public that they should be buying more Perry Como records."

Er, whoa, steady on there, man. You're pulling me a little further than I'd intended to go. As it happened, around that time I bought a Perry Como box set, the usual zillion-CD career retrospective, and, while I like "Chi-Baba Chi-Baba," "N'yot N'yow," "Bibbidi Bobbidi Boo," "Zing Zing Zoom Zoom," "Chincherinchee," "Papaya Mama" and "Hot Diggety (Dog Ziggety)" as much as the next fellow, the charm of the post-war "novelty song" begins to pall after the first 60 or 70.

Unlike Sinatra, Como was happy to sing whatever was put in front of him. On the other hand, "If I Loved You" is just gorgeous, and there's that wonderful, shimmering, ethereal version of "Toyland" on *The Perry Como Christmas Album,* and later on came "And I Love You So" and "It's Impossible," which, even with the soupy Seventies soft-rock arrangements, are better than anything else going on in the Top 40 back then. Perry had a glorious voice but he didn't belabour the point. For one small example, compare Ezio Pinza's "Some Enchanted Evening" on the *South Pacific* cast album with Perry's version: he hits high notes pianissimo rather than fortissimo, which is actually a lot more difficult, and a lot more sensual. The big bellowers like Barbra and Céline use high notes to advertise themselves; Perry was content to serve the song, the mood, the moment.

Still, my original defence of Perry was only in response to the casual slur of some rock bore: horrified to find out that Mick Jagger was not after all a "Street Fighting Man" but only a middle-class art college poseur who'd rather swank about with Princess Margaret, the critic Sean O'Hagen had sniffed, "The Stones are now about as dangerous as Perry Como." When you think about it, the requirement of a pop star to be "dangerous" is a curious one. Fortunately, Puffy, Snoop, the late Tupac and Notorious B.I.G. have been happy to rise to O'Hagen's challenge, and been widely praised for it by his fellow sissy-boy critics at *The New York Times* et al and by such leading thinkers as Harvard Professor Henry Louis Gates.

And therein lies the paradox. These days, it's safe to be dangerous. The really dangerous thing is to be safe, like Perry - wear a cardigan, sit in a rocking chair, sing "It's Beginning To Look A Lot Like Christmas." That's dangerous, or - to use another favourite shaft in the critical quiver – "transgressive." In an age of dreary boomer hegemony, when Al Gore quotes Dylan and Lennon and Liberal hit man Warren Kinsella takes time off from hailing Jean Chrétien's canny stewardship of Quebec golf courses to pen a fond eulogy for the late

Joey Ramone, Perry Como is on the cutting-edge of the counter-counter-culture. In a world of stultifying rock conformism, square is the new cool.

Perry died last weekend. The media didn't make a big deal about it. He had a bunch of Number Ones, sold more records than Dylan, was at one point the highest-paid performer in the history of American TV, but none of it seemed to count for much by the end. As he put it, "I was a barber. Since then I've been a singer. That's it." And the rock crowd had little use for either his singing or his haircuts.

But I only hope I live long enough to see the grizzled boomer rockosauruses fall into the same obscurity. The Sixties Jurassic rockers were the pioneers of the Clinto-Blairite Third Way: socially liberal but fiscally conservative, in favour of low taxes and economic growth, but relaxed about pot and sex and full of bullshit on the environment. No one figured out that winning formula faster than the corporate rockers. The Stones liked the songs of those old-time bluesmen, but they weren't going to end up sitting on a porch in the Mississippi Delta waiting for the welfare cheque. They were the first band with a registered trademark and a merchandising operation.

So now we have rock'n'roll government - bloated, decadent, dependent on lasers and dry ice to conceal its fundamental banality. Those on the right have a choice. They can join the Gores and Kinsellas playing air guitar, as, say, William Weld, former Republican Governor of Massachusetts, used to do. Weld was not only a Jerry Garcia fan but also a great admirer of the Violent Femmes. (They're apparently a rock band and not Weld's core pro-choice constituency.) Or they can be Mister Squaresville like Weld's successor Paul Cellucci. At the state Republican convention, he entered to the strains of the Dave Matthews Band's "Ants Marching."

Alas, whatever advantage he might have gained by this was thrown away when he then told stunned staffers that he'd ...never heard of the song! With aides still reeling from this shock, it then emerged that while attending ex officio the Eric Clapton gig Cellucci misheard the drugs classic "Cocaine" as a song about "Propane." *The*

Boston Herald dubbed him "terminally unhip" and even his spin doctors gave up spinning. With a candour rare in his profession, Cellucci's shell-shocked campaign manager Rob Gray conceded: "He just has bad musical tastes." How bad? The Governor likes ...*showtunes*. The poor guy is now the new US Ambassador to Canada, and Kinsella will be wasting his time if he expects the guy to join in any Sussex Drive Ramonalongs.

But I say: Hail Cellucci! Not knowing the words to "Cocaine," not recognizing the Dave Matthews Band, not having a clue who the Violent Femmes are, these are the signs of a free spirit in today's homogenized culture. One of the best things about George W Bush is that he's culturally conservative and unembarrassed about it - he has the confidence not to be hip, a crucial quality when you're being bombarded with modish nonsense on everything from the environment to hate crimes. Mrs Thatcher was exemplary in this regard. Asked to name a record she liked, she'd cite "How Much Is That Doggie In The Window?" - a title which neatly encapsulates how the conservative's compassion is tempered by cost-benefit analysis.

In other words, despite my sheepishness at the time, I now agree with Joe Queenan's mocking reductio of my argument: Go out and buy more Perry Como records. It's important for conservatives to think outside the box - the box in this case being a 12-CD set of "200 Baby Boom Rock Classics You Hear Every Time You Switch On The Radio." True conservatives don't just talk the talk, they walk the walk. They talk about low taxes and small government, but then they walk over to the CD player and put on "Papa Loves Mambo."

LARGE INTESTINE

Raw sewage

June 21st 2003
The Spectator

O N MONDAY *The Daily Telegraph* gave a big chunk of its Comment Page real estate to Mr Will Day of something called Care International. I've never heard of it myself but doubtless that's because I'm a fully paid up member of Don't-Care Unilateral. Anyway, the headline on Mr Day's column read: "Things Are Getting Worse In Iraq, So Give The UN A Chance."

You can guess how it went from there: "Something in Iraq is going fundamentally wrong... Did the coalition planners think things through..? They can't have... Nobody is safe... complete breakdown in security... making their lives a misery... almost total lack of basic services..."

I didn't see any of this during my stay in western and northern Iraq last month. I went to quite a lot of effort to look for complete security breakdowns, miserable lives and total lack of basic services, but I couldn't find them in Tikrit or Rutba, Kirkuk or Ramadi, Samarra or even Fallujah, where they've had one or two little local difficulties but nothing like the widespread civic collapse Mr Day confidently asserts. Perhaps he's referring to different parts of Iraq. Hard to know, since he cites no specific examples and, indeed, names no Iraqi cities apart from the capital. He includes one hard statistic, in the midst of a paragraph claiming that Iraq is "on its knees. The supply of electricity is erratic and unreliable, clean water is fast becoming scarce and rubbish is piling high in streets flooded by sewage - an estimated 500,000 tonnes of raw sewage, at least, is being poured into the river daily. In the soaring summer temperatures, this is a recipe for disaster. How long will it be before we see this contamination seriously affect the health of the population?"

This passage rang a bell with a correspondent of mine, Nicholas Hallam, who sent me the following press release from the self-same Care International:

> *Sewage treatment has collapsed, resulting in 500,000 tons of raw sewage being discharged into water sources every day... Electricity, essential for many services and previously enjoyed by the remotest villages, is now generally available for less than 12 hours per day in many parts of Iraq. This has an obvious impact on water quantity and quality, sewage treatment, health facilities, education and overall quality of life...*

That was Care International's assessment of the situation in Iraq on January 31st this year, at least according to Margaret Hassan, the director of the Baghdad office, in her testimony in New York before a bunch of UN bigshots. Mr Day's *Telegraph* column of June 16th cheerfully recycles his colleague's January press release, differing only in the root cause of the problem: now, instead of UN sanctions being to blame, it's the American administration. Other than that slight modification, however, far from the headline's claim that "Things Are Getting Worse In Iraq", things seem to be pretty stable. In January, there were 500,000 tons of raw sewage. By June, there were 500,000 tonnes of raw sewage.

As far as I remember, a ton is either just a wee bit more than a tonne or just a wee bit less. But a little research never hurts anyone, so I looked it up and it turns out I'm both right. A British ton is a wee bit more than a metric tonne but an American ton is a wee bit less than a metric tonne. So, giving Ms Hassan the benefit of the doubt, let's assume she was speaking in US tons. That would mean there'd been roughly a ten per cent increase in the amount of raw sewage from 500,000 tons to 500,000 tonnes.

But you know what? When you keep seeing the same big fat awfully round number and only the unit of measurement varies, you can't help feeling that, whether in metric or American or Imperial, nobody at Care International has a clue about how much raw sewage is being pumped into Iraq's water sources. And while I wouldn't want to find 500,000 tons or tonnes dumped in my favourite swimming hole I wonder what exactly the public health consequences are of putting it in the water network of a country the size of Iraq. For purposes of comparison, the Chinese city of Chongqing puts a million tons of raw sewage into the Yangtse each day. In the year 2000, China put 23.5 billion tons of sewage into the Yangtse – that's 63 million tons a day. In Mexico, 100 million tons of raw sewage are said to flow down the Rio Grande every day. Mexico, like Iraq, is pretty hot. My advice to the NGOs is that these big round Iraqi scare numbers need to be rounded up a lot more. 500,000 tons is chickenfeed or, in this case, chickenshit: it's like that moment in *Austin Powers* when the newly defrosted Dr Evil threatens to destroy the world unless it pays him a ransom of one million dollars, and everyone at the UN laughs. You're thinking too small, you Care guys. If half a million tons of raw sewage per day was hitting the Tigris in January, now that the Yanks have been there a couple of months it should surely be at least half a billion.

I can only report my own interaction with Iraq's drinking supply. I crossed into the country from Jordan and stopped for lunch in the first town I came to – Rutba – and the first thing that happened was that the young slip of a lad plunked down on my table a grubby plastic thermos of water with an encrusted spigot and a metal goblet. He seemed to be hanging around and I didn't want him to think I was some sort of NGO nancy boy so I filled up the goblet and drained me a skinful. Not as good as the stuff from my well in New Hampshire, but better than municipal water in Montreal - or in London, where I believe it's mostly recycled Welsh urine these days. This ritual repeated itself across the country and by the third day I was a dab hand, ostentatiously tasting the water and then sniffing to the sommelier: "Hmm. I was hoping for a *soupçon* more coliform bacteria and a rather more playful parasitic worm. Oh, and stick a cocktail umbrella in the human fecal material..." But everywhere I went I drank the water and, aside from mild side effects like feeling even more right-wing than before, I'm fine and dandy.

Others will have their own experience. Robert Fisk of *The Independent* bought 25 loo rolls at the beginning of the war, evidently anticipating a far more gruelling campaign intestinally speaking. But let me make a prediction. However much raw sewage is actually getting pumped into the Tigris and Euphrates, it's never going to be enough to cause a genuinely widespread public health crisis – no matter how much Will Day would like one. "How long will it be before we see this contamination seriously affect the health of the population?"

Seriously? Never.

I would be interested to know, by the way, which streets where are actually "flooded" by sewage.

After I wrote about my trip to Iraq in *The Sunday Telegraph* and its sister papers, I received quite a few e-mails from US troops in the country, the gist of which was summed up by one guy with a Civil Affairs unit near Baghdad: "I'm glad to hear somebody report what's really going on.... the fact that there isn't anything going on." I saw no anarchy, no significant anti-US hostility, and no hospitals at anything like capacity. In other words, I was unable to find Will Day's Iraq. I don't honestly think it exists outside his head: as Dinah Washington once sang, "Water difference a Day makes"; he has miraculously transformed Iraqi water into whine.

But these are the times we live in. There have always been issues on which the differences are so huge they're beyond discussion: generally speaking, it's not worth an American and a European getting into a dinner-party debate over the Israeli/Palestinian question; neither is ever going to change the other's mind because the shared assumptions necessary to engage in argument don't exist. The problem now is that the Israeli/Palestinian template has spread to whole other areas, not least Iraq and the war on terror. In October 2001 Faizal Aqtub Siddiqi, President-General of the International Muslims Organization, said bombing Afghanistan would create a thousand bin Ladens. It didn't. In March this year President Mubarak of Egypt said bombing Iraq would create a hundred bin Ladens. So right there you've got a tenfold decrease in the bin Laden creation program. But even that modest revised target wasn't met. There's widespread starvation and disease and millions of refugees in Iraq. Except there isn't. The Baghdad museum was looted of its treasures. Only it wasn't.

What all these fictions have in common is the prejudice behind them: the article of blind faith that the Americans are blundering idiots who know nothing of the world. It was this that led Robert Fisk, whom my colleague Stephen Glover regards as a "genius", to suggest in print that when the Yanks claimed to be at Baghdad International Airport they'd in fact wandered by accident on to an abandoned RAF airfield. Nobody who knows anything about a modern military or even the kind of GPS technology Chevrolet now include in their mid-price trucks and SUVs would say anything so stupid in print – unless he were so blinded by irrational Yankophobia that he was impervious to anything so prosaic as reality. Likewise, *The Guardian*'s "Gotcha!" scoop in which they brayed that Paul Wolfowitz had finally fessed up: the Iraq war was "all about oil". *The Guardian* was forced to back down when it was pointed out that all Wolfowitz had done was to observe that America had economic leverage against North Korea it didn't have against Iraq, because the latter "floats on a sea of oil".

But, as with Fisk, it's the headlong stupidity that startles. Anyone who's spent even a few minutes listening to Wolfowitz knows that he's actually quite soft-spoken and tonally benign, and that he's a thoughtful fellow who has better contacts among Iraqi political groups than any other western politician. But the Deputy Defence Secretary's sinister reputation depends on little other than the fact that his name starts with a ferocious animal and ends Jewishly. (Christopher Hitchens noted the other day the curious habit of BBC correspondents of referring to the fellow as "Vulfervitz", declining to accept the bearer's own pronunciation of his name.)

The honourable exception among this company is my colleague Matthew Parris, who beginning some months ago declined to sign on to any of the bogus objections of the anti-war crowd - the millions of civilian deaths, etc. Matthew cut to the chase: he was against the war because he thought the real issue was American power in the world. Fair enough. If you believe that, don't duck behind non-existent rubbish like rampant cholera and all-about-oil fantasies. Last week, Matthew said that, had he been President, he would not have invaded. That way, "international law would not have been violated, swollen-headed neocons would not have gained sway, the yee-hah tendency in US foreign policy would have been restrained, precedents for future unilateral regime-changes would not have been set, Nato would be intact, the UN Security Council would not have been damaged, America's relationship with Europe would have remained good, and Britain would still be on speaking terms with our EU partners."

Actually, aside from anything else, they're all reasons why I was in favour of war. If the overriding issue for M Parris is American hegemony, the issue for me is the rise of transnational neo-imperialism. I'd rather take my chances with nation states and great power politics than submit to "international law". I think Nato and the UN Security Council need "damaging", and so does America's relationship with "Europe". And the jetset humanitarians, as represented by Will Day, might also benefit from being forced to rethink their act. There is, of course, a real humanitarian crisis in the world today – in the Congo, an environment blessedly free of blundering Yanks, where "international law" has ridden to the rescue and, as in the Balkans and elsewhere, the UN is providing the usual genteel multilateral cover for ethnic slaughter. But, because it doesn't accord with the New Universal Theory of Texan-Zionist neocon aggression, nobody cares.

In Iraq, the Americans and British are muddling through; in the Congo, "international law", as represented by the French and the UN, is failing big time. That's my view and it happens to fit my prejudices. But it also fits the facts.

269

ASS

Wide shot

September 25th 1999
The Daily Telegraph

SAY WHAT YOU like about this US presidential campaign but the third-party candidates have never looked better. Last time round, back in '96, we had to make do with Ross Perot, the diminutive jug-eared goofball who claims to have survived assassination attempts by the Viet Cong, the Black Panthers and the father of the guy who plays Woody in "Cheers". In those days, short of getting your dad to whack Perot, the average sitcom star played no part in presidential politics. But now Cybill Shepherd, star of the archetypally average sitcom "Cybill", has declared that, like Warren Beatty and Donald Trump, she's considering a run for the White House. And why not? Parlaying a sitcom into a presidency doesn't seem so far-fetched when you consider Bill Clinton's managed to turn a presidency into a sitcom.

The creamy-skinned straw-blonde southern belle's main issue is abortion rights, though it's not known how impressive her campaign chest is. Incidentally, by "impressive campaign chest", I'm referring to her financial backing and not the topless scene in *The Last Picture Show*, impressive though that was. While Ronald Reagan ran under the slogan "You Ain't Seen Nothin' Yet!", President-elect Shepherd's would more likely be: "There's Nothin' You Ain't Seen Yet!"

Not everyone is so impressed, though. In her book *You'll Never Eat Lunch In This Town Again*, Hollywood producer Julia Phillips dismissed Cybill as someone "with little talent and a big ass." But you could say the same about Clinton and it didn't stop him. Indeed, Cyb and Bill have much in common, not least a fondness for Elvis. In her forthcoming memoir, *Cybill Disobedience*, Miss Shepherd tells for the first time how a drugged Elvis fell asleep on her *in flagrante delicto*. This is late Elvis, of course: He was at that stage in his life where come 10pm he'd just put the catsuit out for the night and go to bed. Revealing that her partners doze off in the middle of sex with her could seriously erode Al Gore's core support among narcoleptics.

Otherwise, her editor at HarperCollins says the sex scenes in her book are better than in any other presidential candidate's memoirs. While the other guys just drone on about their record in Congress, Cybill relates in detail her record in congress. "He was sexually conservative," she says of Elvis, "trapped in a macho thing." (Except on the night he fell asleep halfway through, when the macho thing was trapped in her.) I can't remember (and possibly neither can he) whether she nailed Warren Beatty back in the Seventies, but I'd have thought, statistically, there's a good chance Cybill and Warren are the first two presidential rivals to have had sex with each other. Make that three: Donald Trump's dated a lot of ex-models in his time.

Sorry to bang on about sex, but that's mostly what Cybill talks about: "Men hit a certain age, they don't want to do it anymore. But women never stop wanting to do it." She's spokeswoman for the "Say Yes To Midlife!" campaign: "We're dealing with the last frontier. We're reinventing menopause and blowing all the myths to bits," she says. "For me, sex didn't just get better - it got great!" She devoted two episodes of her sitcom to the menopause. After that, CBS cancelled it. Still, she's determined to be the first world leader to bring up "reinventing menopause" at a G7 summit.

It's sad that those uptight suits at the network couldn't handle a strong, sexual, mature woman in a sitcom, but fortunately Cybill has a second string to her bow. She sings. Ever since her first album, the aptly named *Cybill Does It - To Cole Porter!*, she's sung standards at anyone who'll listen. In Britain, Radio 2, under the impression that she was some sort of huge singing star in America, offered her a staggering amount of money to fly over and sing with the BBC Concert Orchestra: Proportionately, they've probably made a bigger financial contribution to her campaign than the Chinese Politburo did to Clinton's. But no doubt, as with the Reds, it'll pay off in the long run. Either way, Cybill is the first singing thespian to run for president since Stuart Hamblen, a cowboy extra and songwriter (he composed "This Ole House", a Number One for Rosemary Clooney, and later Shakin' Stevens), ran on the Prohibition Party ticket in 1952.

These days Cybill isn't for prohibition of any kind - least of all on abortion. Democratic rivals Al Gore and Bill Bradley are both "pro-choice", as the preferred circumlocution has it, but only Cybill is pro-abortion. She loves it. "This is the happiest day of my life," she said at one abortion rally, "except for the days my children were born." Has she got any other issues? Well, yes. She's concerned about Supreme Court judges …who are insufficiently pro-abortion. She's extremely worried about terrorism …at abortion clinics. As for East Timor, well, unless there's a foetus in there somewhere, don't expect a policy position any time soon.

But don't underestimate her either. Cybill's an experienced campaigner: She played a Senate campaign worker in *Taxi Driver,* the film where Julia Phillips tried to persuade Martin Scorsese to take out the wide shot of her butt. "I'm Italian," said Scorsese. "I love it!" Italian-American butt-fetishists could be a crucial demographic in states like New York.

The only puzzle about the Shepherd campaign is why *The New York Times* - who after all see eye to eye with her on pretty much everything - sniffily regard her as a joke candidate and refuse to mention her. I'll bet the floundering Gore campaign isn't treating her as a joke. She did it to Cole Porter, she'll do it to Al.

RECTAL APERTURE

Major league

September 7th 2000
The National Post

STAND WELL back. Bush has - gasp! – "gone negative". He has unleashed - aaargh! – "the politics of personal destruction". And what's more, the guy he's personally destroying is a blameless journalist!

As every TV viewer in America knows by now, while up on the podium waiting to speak, genial George Dubya spotted a familiar face in the crowd. Forgetting there was a microphone mere inches from his plastic smile, the Governor chortled to Dick Cheney: "There's Adam Clymer, major-league a****** from *The New York Times.*"

"Oh, yeah, he is," agreed Cheney. "Big time."

By a******, I mean, of course, ass**** or, according to what paper you read, ***hole. At any rate, as *The Washington Times* noted, it was "a vulgar euphemism for a rectal aperture". I would have thought a "vulgar euphemism" was an oxymoron, but in this case the neutral anatomical definition does indeed sound more queasily vivid. "Bush Uses Profanity About Reporter," the Associated Press said. Given that profane means "serving to debase or defile what is holy", I'm not sure what the AP's getting at here. To the best of my knowledge, no major religion regards the rectal aperture as sacred, except possibly adherents at the shrine of *The New York Times*.

But then, in America today maybe it's profane to utter any criticism of the holier-than-thou media. Bryant Gumbel, host of the widely unwatched CBS "Early Show", seems to think so: "Bush may have even taken yet another step backwards by sticking his foot in his mouth with a vulgar comment." On this issue, Bryant knows whereof he speaks. The grand panjandrum recently interviewed a Christian conservative about the Supreme Court decision on gay scoutmasters, at the end of which he handed over to the weather guy. Unfortunately, the camera lingered on him just long enough to catch him dismissing his interviewee as a "f***ing idiot".

But it seems it's one thing for a media [expletive] to call a conservative an [expletive], quite another for a media [expletive] to be characterized as such by a conservative [expletive]. The expletive in question, Adam Clymer, is after all the acclaimed author of *Edward M Kennedy: A Biography*, in which he salutes his subject "not just as the leading senator of his time, but as one of the greats in history… a lawmaker of skill, experience and purpose rarely surpassed since 1789".

Not that Clymer, schooled in the art of *Times* impartiality, doesn't tackle the debit side of the Kennedy ledger: his "achievements as a senator have towered over his time, changing the lives of far more Americans than remember the name Mary Jo Kopechne". Oh, well, that's okay then.

I don't know how many lives the Senator's changed - he certainly changed Mary Jo's - but I'm struck less by the precise arithmetic than by the curious equation: How many changed lives justify leaving Miss Kopechne struggling for breath for hours pressed up against the window in a small, shrinking air pocket in Teddy's car? If the Senator had managed to change the lives of even more Americans, would it have been okay to leave a couple more broads down there? Such a comparison doesn't automatically make its writer an a******, but it certainly gives one a commanding lead in the preliminary qualifying round.

But the accuracy or otherwise of the Governor's observations on Mr Clymer's assholian status are of no concern to his fellow media types. Instead, on the news shows, the clip of Dubya's frightful *lese-majesté* has been broadcast again and again. The benign explanation is that, well, you know how the networks tend to play up anything they get on video. But there's all kinds of stuff they get on video that mysteriously never makes it to air. While I was mooching about the

Democratic Convention last month, six Boy Scouts walked out on stage to lead the Pledge of Allegiance. They were booed by delegates. "It was pretty insensitive," said one Dem - not about the booing, but about the decision to let the Scouts appear at the Convention, in the wake of the Supreme Court ruling permitting the organization to exclude openly gay Scoutmasters. So there was the rank and file of the Democratic Party jeering at half a dozen bewildered young boys. This is amazing, I thought. I'd already filed my column for the day, but I felt sure ABC, CBS, NBC et al would be on top of the issue. But no, not a peep. *The New York Times* and *Washington Post* likewise considered it unworthy of mention.

By contrast, when gay Congressman Jim Kolbe addressed the Republican Convention, the *Times* devoted no fewer than four stories to his reception: As he spoke, 12 members of the Texas delegation, being disapprovers of homosexuality, quietly bowed their heads in prayer. Even worse, most of the rest of those hard-hearted Republicans "offered only tepid applause," tutted the *Times*. Tepid applause! Is there no end to the hate these bigots aren't prepared to spew? Unlike the loud boos from the Dems, the silent prayer and tepid applause at the GOP were gone over at length by the media.

US newsrooms are fully committed to diversity of gender, diversity of race, diversity of orientation, diversity of everything except the only diversity that matters: diversity of thought. 92 per cent of journalists voted for Clinton-Gore. I'm not joking: that's the statistic. That doesn't mean 92 per cent of them are a******s, but it does help explain why their papers are the dullest and most unreadable in the English-speaking world. Conformity is their watchword. They're all agreed on what news is: news is "tepid applause" from Republican homophobes, not boos and jeers from tolerant Democrats. When political coverage is a private club, it's no wonder more and more of the public just leaves 'em to get on with it.

William Powers summed it up beautifully in the *National Journal*:

The journalistic establishment is like one big, pretentious snot-nosed French waiter, and it's time for America to hurl a glass of ice water in its face and give it the boot.

Calling 'em a******s is an excellent start. Way to go, Dub!

GROIN

Strain

November 29th 1997
The Spectator

SEYMOUR STREET and Seymour Hersh are several thousand miles apart, except when it comes to evaluating presidencies. The latter is a Pulitzer prize-winning author, the former is a thoroughfare in downtown Vancouver, where President Bill Clinton was for this week's Asia-Pacific summit. I don't know whether his motorcade passed down the aforementioned street but, if it did, he might have noticed the marquee above the Penthouse strip club: "Welcome Prez Clinton, our lips are sealed" - ie, he and the girls are old pals.

It could be. If you'll forgive a little maple pride, many of the most highly regarded strippers in the United States are Canadian, their whirling tassels extending even unto Little Rock, Arkansas.

But, on balance, I think the ladies were just having a joke. Though the President is said to be fretting about his place in history, on Seymour Street and elsewhere it's already secure: he's one of the great comic figures of the age. South of the border and back on the East Coast, Seymour Hersh can only marvel. He's just published a book about President Kennedy, who, even by the most tactful underestimate, had a sexual appetite that makes Bill Clinton look like Marie Osmond. "You know," JFK remarks to Bobby Baker, secretary to the Senate Democrats, in the Oval Office one morning, "I get a migraine headache if I don't get a strange piece of ass every day."

And it seems he did, even in the White House: society gals, hookers, East German Communists, three-in-a-bed, two-in-a-bath, with a Secret Service agent standing by to shove the girl's head underwater at a given signal, thereby causing vaginal contractions and thus intensifying the Presidential orgasm. All ass, all the time, regardless of whether it was a quiet day with not much going on or the height of the Cuban missile crisis.

Hersh is no right-wing, Kennedy-hating kook. As the man who alerted America to the My Lai massacre and CIA domestic spying, he has impeccable liberal credentials. The sex is in there only insofar as it impinges on national security, defence contracts and White House operations, and it's well-sourced - the whores and nude pics are confirmed by the gallery owner who framed White House photos for over three decades and by various Secret Service agents - all with names, potted biographies and even photographs. Yet *The Dark Side Of Camelot* has been denounced as "evil" and "utterly without credibility" by the likes of *Time*, *The Washington Post*, *The New York Times* and even *The Spectator*'s High Life Correspondent.

Camelot's "fleeting wisp of glory", its "one brief shining moment" is proving surprisingly durable: after 35 years Kennedy's vast army of sycophants is as fanatically loyal, as ruthlessly protective as ever. What are they so steamed up about? Anyone who investigates the Kennedy White House comes to pretty much the same conclusions as Hersh. The British author Nigel Hamilton wrote a cracking account of Jack's early years, *JFK: Reckless Youth*, the first part of a two-volume biography. The second half never appeared. Hamilton, a Kennedy admirer, was so disgusted by what he subsequently uncovered about his hero that he abandoned the project. That's what so riles Hersh's detractors: this is all the stuff they were too dazzled to spot at the time.

Why is it that Seymour Hersh and even Seymour Street see more than the Washington media? Well, for one thing, most of the press don't want to see: the grandees of American journalism have a lofty view of their own profession and therefore extend it naturally to the profession they spend most of their time covering. Among the most vociferous trashers of this "evil book" is Hugh Sidey, the veteran *Time* White House correspondent. So it comes as a surprise to discover that he's one of Hersh's principal sources. Presumably, he thought he was contributing to quite a different kind of book. At one point, he relates a meeting he had in the Oval Office with Kennedy: "He looked up at me and says, 'Come on, Sidey. Let's go swimming.' I said, 'Mr President, that's the one piece of equipment I never thought to bring when I came over for an interview.' He said, 'Oh, in this pool you don't need a suit.'"

Kennedy and Sidey head out to the White House pool, at which point the reporter finds himself facing one of those awkward questions of etiquette on which Miss Manners is silent: "I'm confronted with this problem of who removes his trousers first - the President or the guest?"

But you've got to be quick on the drawers to drop 'em faster than JFK. "Obviously a man of practice," chuckles Sidey - and soon interviewer and interviewee are splashing away like a nude synchronised swimming team. You couldn't ask for a better image of American political journalism: two members of the same cosy club gliding along side by side.

Needless to say, whatever distinguishing characteristics the President might have had, *Time* readers didn't get to hear of them. The American press likes to scoff about Britain in the Thirties, when, if you wanted to read about the Prince of Wales and Mrs Simpson, you had to buy a US paper. Today, the Clinton Presidency has reversed the process. "I've just got back from London," Pat Buchanan told me a couple of years ago. "I can't believe the stuff you guys are running about Paula Jones." "They put that in the papers?" said Eleanor Clift, doyenne of *Newsweek* and Hillary confidante. Not in the kind of papers Miss Clift reads.

History repeats itself - first tragedy, now farce. First, the Broadway company of *Camelot*, now the touring production of *When Did You Last See Your Trousers?* For most veterans of the *Camelot* era, the journey from Kennedy idealism ("One Brief Shining Moment") to Clinton discussing his underwear on MTV ("One's Momentous Shining Briefs") undoubtedly marks a precipitous decline in presidential glamour. But it makes the enduring prissiness of the

Washington press seem even more ludicrous. Why should they be so concerned about the dignity of political office when the officeholders themselves aren't? On the eve of last month's gubernatorial election in New Jersey, Governor Christie Whitman went on the Howard Stern radio show. He congratulated her on cutting taxes, her pro-abortion stance and - last, but by no means least - her "fantastic breasts". Thank you, said the Governor. Unconcerned about motor insurance rates (the dominant issue in New Jersey), Howard was interested to know the state of Mrs Whitman's sex life. It was especially good at weekends, she said, when they were at their country place. Howard wanted her to know what a terrific body she had. "Thank you again," she said. He then shared with the tri-state listening audience some thoughts on what it would be like to "do" the Governor.

In such a world, it's not surprising that the Washington press corps now regards itself as the dignified part of the constitution. But, even so, the resilience of Camelot, in the face of all the evidence, is one of the marvels of the modern age. As the slogan puts it, "If you loved his style, you'll love JFK PT wear". PT stands for "Patrol Torpedo", a reference to JFK's Second World War boat, PT 109. In this exciting new range of leisurewear by Kerry McCarthy, the President's cousin once removed (though it seems no Kennedy cousin is ever really removed), the clothes are not based on what the well-dressed PT crew member is wearing when the Japs slice his boat in two: that, after all, was one of the few moments in his life when JFK was not thinking about image. Instead, it's the sort of yachtwear - JFK cap, $19.95 - the President was wont to favour when mooching about off the Hyannis coast in less turbulent waters. Each item in this exclusive collection, however, features a reproduction of the "PT 109" insignia which the young Kennedy sent over to Kerry McCarthy's mom during the war. "Now you too can wear John Fitzgerald Kennedy's PT patch," says the advertisement. "You too can be part of that Kennedy style." And the Kennedy style can still drown out the Kennedy satyriasis, the Kennedy pill-popping, and the Kennedy underwater orgasm intensification.

If nothing else, Hersh's book and the sledgehammer condemnation of it raises the question of just how far the Washington media are prepared to take their lack of curiosity. Hersh steers clear of Kennedy's assassination, except for one passing observation. In September 1963, "while frolicking poolside with one of his sexual partners", the President severely tore a groin muscle and was prescribed a stiff canvas shoulder-to-groin brace that locked his body upright. "Those braces," writes Hersh, "made it impossible for the President to bend in reflex when he was struck in the neck by the bullet fired by Lee Harvey Oswald. Oswald's first successful shot was not necessarily fatal, but the President remained upright - and an excellent target for the second, fatal blow to the head." It's an interesting theory, borne out by the available footage. If you talk to Kennedy insiders, they say he'd worn a brace on and off for decades due to chronic back pain. On the other hand, his lifestyle wasn't exactly conducive to mitigating back pain.

"Kennedy may have paid the ultimate price," Hersh concludes, "for his sexual excesses and compulsiveness." You don't have to swallow that whole to reflect on the broader possibilities. In all the flights of fancy the grassy knoll has spawned - was it the CIA? the FBI? LBJ? - for 34 years there's been a weird reluctance to look for an answer in the pathological recklessness of Kennedy himself.

WOMB

Stork report

October 31st 1999
The Sunday Telegraph

A FEW YEARS back, I happened to be on a radio show with Gore Vidal, who, for some reason, assumed I was gay. During the off-air banter, another guest began talking about her newborn. The great man of letters gave me a conspiratorial twinkle and sighed wearily, "Breeders!"

Gore may still be a non-breeder, but he's the last of a dying non-breed. The 21st century is upon us, everyone's broody, and, in case Gore's wondering, it no longer involves anything as ghastly as being in the same room as a woman with no clothes on. The stork has diversified: You no longer have to look for his little bundles of joy under the gooseberry bush - you can order them online. And, as his traditional market - or, as *The Guardian* calls it, "the family lobby" - has shrunk, he's moved on to expand his share of key niche demographics. Now, infertile women can have babies, and so can sexagenarian women and gay men.

Take Barrie and Tony, a couple from Chelmsford, Essex. They'd been trying for a baby for some time, but nothing seemed to work. Then it occurred to them that this might be because they're both men. So they found a woman who happened to have four eggs lying around that she hadn't yet auctioned over the Internet. You would think two boys and a girl would have been enough, but they figured they needed someone else just to even up the numbers, so they roped in another woman who happened to have a rare nine-month vacancy in her fallopian timeshare. After that, it was just a question of getting the girls in the mood: the lights down low, Johnny Mathis on the hi-fi, the FedEx package with Barrie and Tony's beaker of co-mingled sperm on the coffee table, the cheque for $200,000 in the mail, and the turkey baster wandering in from the kitchen with a come-hither look in his eye.

The result of this happy union is twins: a boy for him and a girl for him, to modify "Tea For Two". Barrie Drewitt and Tony Barlow are planning to name their son and daughter (or vice-versa) Aspen and Saffron Drewitt-Barlow. In a landmark decision in a California court, the proud parents will be the first British couple to both be named as father on the birth certificate, though neither mother rates a credit. The babies have not yet been born, but both mother and surrogate mother and co-father one and co-father two are doing well: Barrie and Tony still have a few eggs in the freezer from the same woman so, in a year or two, they intend to provide Aspen and Saffron with a sibling named after some other spice or ski resort. "This ruling," said Tony, "affirms that gay couples are entitled to the same fundamental procreative freedoms as heterosexual couples."

It's fair to say heterosexual couples of the old school did not think of "procreative freedom" as an "entitlement" - like, say, public education or a senior citizen's bus pass - but rather as, to use an archaic phrase, a "fact of life". Today, though, there are no "facts of life": de facto, it would seem biologically impossible for Messrs Drewitt and Barlow to come together to produce young Saffron or Aspen, but, de jure, it's a breeze. Neither parent supplied the egg, neither parent carried the child, neither even went to the minimal effort of personally whacking the seed up the ol' vaginal canal, but nonetheless the birth certificate will certify that they're responsible for the birth - essentially for the reason that that's the way they want it: Yes, sirs, that's your baby/No, sirs, we don't mean maybe. "The nuclear family as we know it is evolving," said Barrie. "The emphasis should not be on it being a father and a mother but on loving, nurturing parents, whether that be a single mother or a gay couple living in a committed relationship."

That's great news if you're a gay couple living in a committed relationship or a single mother living in several uncommitted relationships, but in the murky territory in between lurk all kinds of unsuitable parents. In Britain, as was reported last week, Penny and Stephen Greenwood's baby will emerge from the womb and immediately be taken away by social workers and put into foster care. The Greenwoods, of Bradford, who have already had one child confiscated by the state, are both epileptics and, although they insist their conditions are mild and controlled, the authorities aren't prepared to let them be loving, nurturing parents. Apparently, it could be very traumatising for a child to see his parent with his head thrown back and his tongue lolling out - unless, of course, it's at the local gay bathhouse.

Despite the claims of the technobores, in the second half of the century hardly anything has changed - except the nature of change. A young man, propelled by an H G Wells time machine from 1899 to 1949, would be flummoxed at every turn. By contrast, a young man, catapulted from 1949 to 1999, would, on the surface, feel instantly at home: our bathrooms, kitchens, cars, planes and high-rises have barely altered. Instead, having run out of useful things to invent, we've reinvented ourselves and embarked, with a remarkable insouciance, on redefining human identity. In the two decades since the first test-tube baby, "procreative freedoms" have become the new frontier. We began with "a woman's right to choose" - whether or not to abort. Next came an Asian's right to choose the sex of his child and get rid of any unwanted girl foetus (at one point China had 153 boys for every 100 girls). We've now moved on to a couple's right to choose their baby's genetic characteristics on the Internet, a lesbian's right to choose to be impregnated by a gay male friend, and a career woman's right to choose to have her eggs frozen in her late 20s, stored away and fertilised in her 40s or 50s or whenever she feels she's ready to raise a baby. There is a logical progression in all this: if you have the right to end life (with abortion), surely you also have the right to decide when, where, how and with whom you wish to initiate it. And, in one sense, the culture of death and the culture of new life form a kind of balancing act: if there is a gay gene and straight parents start aborting their gay foetuses, it seems only fair to allow gay parenting as a kind of corrective. Likewise, if girl foetuses are shouldering an unfair percentage of abortions in regrettably misogynist societies, female numbers can be kept up by human cloning - which, in theory, eliminates the need for sperm. And, if you don't need sperm, do you really need men? Women could go on cloning women until Amazons ruled the earth, except for a handful of gay male enclaves in West Hollywood and Miami.

Human cloning will happen, if only because there's a market for it - as there's proved to be with eggs and surrogates. If it were simply a matter of wanting to be "loving, nurturing parents", adoption would do it. But there's a biological imperative driving these advances. Since Barrie and Tony are so proud of their "committed relationship", it must be irksome to have to let Tracie the egg-donor and Rosalind the womb-renter into the picture, since neither woman has any commitment to the relationship once the cheque's cleared. Lesbian parents, like the pop star Melissa Etheridge and her partner Julie Cypher, would in future have no need of third parties: the clone of one would grow in the womb of the other - and what could be more loving and nurturing than that? The first human clone will enter the world in a clinic in Mexico or Morocco or some such, but one day she will come to the United States or Britain and endeavour to get a driving licence and at that point, even if cloning remains illegal in those jurisdictions, the state will balk at turning her away because she's officially a non-human. They will recognise her as a legal human being on the grounds that that's what Morocco says she is - just as the British authorities are recognising the California court's decision on Barrie and Tony.

The public will most likely go along with these innovations. Half a century ago, Ingrid Bergman gave birth out of wedlock and it almost finished her career. Now, single mom Jodie Foster is put on the cover of *People* magazine as a paragon of motherhood, and everyone thinks it's bad form to inquire who or where the dad is, never mind whether a woman who thinks the only function of a father is to get the globules of bodily fluid into the beaker and then push off is really such a great role model. As always, the assault on traditional values is more positively expressed as a tolerance for diversity. "There is no one 'perfect' model on which all family structures can be based," Barbra Streisand recently told America's leading gay newspaper, *The Advocate*. "If we surveyed human history, we would see representations of every type of possible social arrangement." Really? Miss Streisand doesn't give any examples, but you could survey all human history and be hard put to find any precedent for Barrie and Tony's arrangement. The truth is that, rather than returning to some pre-Judaeo-Christian utopia, we've chosen out of sheer self-indulgence to embark on a radical rejection of a universal societal unit.

Maybe it will work out. Maybe in 15 years' time Aspen and Saffron will be sitting in class surrounded by offspring of lesbian couples and geriatrics plus a handful of clones, and they'll all be happy and well-adjusted. Or maybe they'll be like America's vast army of children born out of wedlock - statistically, six times more likely to develop drug habits or commit serious crimes. Either way, the "family lobby" can do little about it, except, as in America currently, object to grade-school youngsters having books like *Why Melissa Has Two Mommies* as a set text. In fact, these days, all too many children have two mommies - one "bio mom", one step - plus a couple of dads, mommy's current boyfriend, whatever. It's not Barrie and Tony who destroyed the nuclear family, but heterosexual parents who have trivialised traditional notions of family responsibility over the past 30 years. We cannot save family values from Barrie and Tony because there's nothing left to save.

URINARY TRACT

Best direction

January 29th 1994
The Daily Telegraph

ON RADIO 4 a year or two back, the *Telegraph*'s Brenda Maddox famously flayed P G Wodehouse because neither Bertie nor Jeeves ever visits the lavatory. In other words, they're not real. Well, by that criterion, Brenda should be having a ball at the movies these days. In film after film, it's micturition-a-go-go.

The willingness to relieve oneself on screen is to Hollywood's current leaking, er, leading men what nudity was to Sixties actresses. And, unlike nudity, it's not gratuitous. Entire plots hinge on the P-moment: in *Mrs Doubtfire*, where Robin Williams dresses as a nanny, his identity is rumbled only when one of the kids catches "her" standing over the porcelain. In John Woo's *Hard-Boiled*, the hero's burning clothing is extinguished by a timely cascade of amber. In Clint Eastwood's *Perfect World*, Kevin Costner's quick getaway is imperilled by his boy sidekick's need to empty an extraordinarily full bladder. In John Turturro's *Mac*, the bond between father and son is forged during an al fresco call of nature in which poppa demonstrates that the way to keep your skin looking so young is to skip the Camay and just rub urine on your hands.

Now, while happy to disrobe for cinematic sex scenes (if anyone asked), I'm not sure I'd be prepared to do any of the above. Brenda and I are both North American, but, while Brenda has clearly absorbed the traditional British fascination with such matters, I retain a certain prudishness. I remember how offended an uncle was by a scene in *Fun With Dick And Jane*, in which, after a bank hold-up, Jane Fonda takes the loot into the bathroom, sits on the can and runs the taps in the basin - the discreet giveaway that she's doing more than sitting there. (If I understand this aspect of our complicated bathroom etiquette correctly, on entering the powder room you immediately turn on the faucets so that passers-by will assume you're in there rinsing your socks.) That was only 1977, but, as Miss Fonda might say, "*Après moi le déluge.*"

The present trend dates, I reckon, from Mike Figgis' *Liebestraum* (1991), in which the county sheriff unleashes a torrent down the side of his police car that lasts longer than the title song in *Hello, Dolly!* Surely such a performance should be eligible for a special Oscar - for Best Direction, if nothing else. After all, all these sequences seem to be filmed from such bizarre angles and with such ludicrously splayed legs for no other reason than to afford us a ringside view of as much bodily fluid as possible. In *Mrs Doubtfire*, Robin Williams is exposed mainly because he stands at a weird angle to the toilet, as no man ever would. How do they shoot these scenes? Do they use body doubles and tank 'em up with weak tea ("Take 14, everybody. Let's try and get one in the can")? Or is it a procedure involving hose pipes threaded down trouser legs?

Either way, it's virtually compulsory if you want to pull in that box-office moolah. Or, as Busby Berkeley's tapping chorines would sing, if they remade *Gold Diggers Of 1933*, "Urine, The Money". Back then, if the film sagged a bit in the middle, you'd nip out to the john. Now, when it sags a bit in the middle, the movie itself nips out to the john. Our dramatists have taken Brenda's line too literally: Art is pursuing drab reality to its watery grave.

PENIS

The executive branch

October 19th 1997
The Sunday Telegraph

"IN TERMS OF size, shape, direction, whatever the devious mind wants to concoct, the President is a normal man." Thus, last week, Bill Clinton's lawyer, Bob Bennett, with another landmark first for this Administration: never before has an official statement been made about a Presidential penis.

The distinguishing characteristics of Grover Cleveland, Warren Harding, Herbert Hoover, all are lost in the trouser-folds of history. The founding fathers, in their wisdom, had not foreseen a formal constitutional role for the First Member. That curious innovation has been left to the 42nd President, whose penis has lately had more official engagements than its nominal Commander-in-Chief: on October 3rd, in preparation for Paula Jones' sexual harassment suit, it was formally examined by the National Naval Medical Center's urology chief, Captain Kevin O'Connell, who, according to Mr Bennett, pronounced everything shipshape, metaphorically speaking.

It's hard not to feel a sneaking sympathy for President Clinton as he tries to shake off Mrs Jones - the kind of sympathy one feels for Benny Hill in those speeded-up musical finales, when Benny, the leering predator-turned-prey, finds himself pursued by irate dolly birds and their menfolk. Substitute a high-speed "Hail To The Chief" on the soundtrack and a posse of ambitious lawyers, and that's the Clinton second term in a nutshell. Last week he found himself facing a list of 72 written questions regarding the movements of his organ for most of the 1980s. Whatever its size or shape, Mr Clinton's penis is casting a long shadow over his presidency.

In one sense, one could look at it as a triumphant if belated vindication of the Revolution: Bill Clinton has effortlessly demonstrated that, when it comes to sexual buffoonery, one elected head of state can do the job of the entire Royal Family. On the other hand, his predicament also illustrates the differences between a society with a written constitution and a Bill of Rights and one that proceeds according to informal convention. Under the informal conventions of British society, Bill Clinton would have been exposed in *The Sun* by some vengeful waitress about six weeks into his first term as Governor, and either been hounded from public life or settled down to become a well-loved figure of fun. How enviably straightforward such a system is.

That can't happen in America, where the mainstream media are, according to Clinton critics, part of the culture of complicity: I think they're more like Victorian matrons, aghast at Mr Clinton's rampaging piano leg, and preferring simply to draw a discreet veil over it. No television networks reported Mr Bennett's statement or the President's medical examination, and news of the Jones legal team's submission of explicit questions was confined in respectable papers mostly to wire-service round-ups under the heading "National Briefs". To distract attention from the national briefs, President Clinton spent the week in Latin America. So the whole thing is played out according to the elephantine process of US law, weighing Mrs Jones' grievance against the President's duties, the Bill of Rights against the rights of Bill.

However it turns out, it's already the most expensive (alleged) trouser-drop in history. According to her deposition, it is Mrs Jones' contention that what she saw in that hotel room in 1991 was bent, at an angle, a condition caused by something called Peyronie's disease. If she's right, that would tend to corroborate her story: Peyronie's disease afflicts about one in a hundred men. Gennifer Flowers, who claims to have spent over a decade in close contact with Exhibit A, says she never noticed anything unusual about the curvature. Unfortunately, as the President denies ever having had an affair with Miss Flowers, he can hardly call her as an expert witness. So we will have to make do with Bob Bennett's statement that there is no unusual curvature – or, to modify an earlier Presidential denial, "It is not a crook."

It's ludicrous, it's degrading to the presidency, but it's the fault of the American people. In 1992 they decided to overlook Mr Clinton's, ah, indiscretions. The law, alas, has less discretion. Janet Reno, the Attorney-General, should know: as a prosecutor in Florida, she was a tireless crusader against sexual abuse, throwing one dismal nonentity after another into jail on the flimsiest of anecdotal evidence. The liberal elite who crafted such laws should have foreseen that one day they would be turned against their own.

Even if you discount all the stories of Clinton's alarming sexual adventures, his presidency has nevertheless found its natural level: in the most Sophoclean way, it's a rendezvous with destiny. Throughout his political life, Mr Clinton has been almost pathologically unable to avoid drawing attention to his private parts. He has discussed his fondness for both Y-fronts and boxer shorts on MTV. For years he made it a practice to give away his underwear and claim it as a deductible item against his income tax - in 1986, $2 per pair for three pairs of used briefs; in 1988, $15 for a pair of old long johns. For his entire career, the President has flown by the seat of his pants.

If Mrs Jones has her way, the bottom is about to drop out.

GLANS

Piercing analysis

November 22nd 1990
The Independent

IN THESE difficult economic times, here is a heartening British success story: *Skin Two*, a publication for devotees of sado-masochism, has increased its circulation from 8,000 to 30,000. "That's more than *The Listener* and only a bit less than *Record Mirror*," pointed out Mark Chase on Channel 4's "Sex Talk" in what was intended to be a startling comparison but sounded more like an all too plausible descent into ever more esoteric debauchery: from colour posters of Inspiral Carpets to whip-wielding thigh-booted dominatrices to the ultimate degradation of long, lonely nights spent reading transcripts of the Reith Lectures.

Many of us today, following the example of the Conservative leadership election, are in the mood for a bit of S&M or, alternatively, infantilism. These, according to Peter Boyd-Maclean's fetish documentary, are the two extremes of threat and security, the basic impulses which underlie our sexual needs, and indeed seem to encompass the Tory benches' complicated relationship with Mrs Thatcher. On the one hand, Madame Whiplash; on the other, Mummy Hazel of the Hush-a-Bye Baby Club, who for a small fee will take you to the nursery, put you in your best bib and nappy, spoon-feed you in the high-chair, tuck you in your cot and then read a bedtime story. "Are you still nice and dry?" she coos to the pudgy middle-aged bloke in the Xtra-large Pampers. "There's a good baby."

Sympathetically lit, the nursery bathed in radiant autumnal sunlight, Mummy Hazel's fluffy cocoon was a not unappealing prospect at the end of a long day. Indeed, it was only when the film

292

ended that I realised I was sucking my thumb and dribbling. The "adult baby scene" seemed sweet, charming, harmless - everything, in fact, but what the presenter had billed it as: "an exciting alternative to penetrative sex". Like the rest of the programme, it had more going for it as an exciting alternative to stamp-collecting or model airplane building.

The overall air of the documentary was that of a spirited village fete where the vicar had replaced the egg-and-spoon race with some less familiar but even jollier games. There was a sort of horse-and-buggy event where the buggies were occupied by masterful types with riding-crops and the horses were replaced by sex slaves in black rubber. And even the genital piercer sounded like one of those slightly obscure village artisans Brian Johnston used to interview on "Down Your Way":

"The nicest of the male piercings," said the hale fellow well metalled, "is what we call the Prince Albert, which is where we put a ring through the end of the urethra and it comes out just behind the glans itself at the base. This is a super little piercing."

Was it my imagination or did I hear just the slightest dull metallic thud from the direction of his own zipper? Apparently so. "It's in fact," he said, beaming with pride, "my own piercing."

This raises all manner of interesting questions, and not just why is it named after Prince Albert? (He is rumoured to have worn one, and to have proved his metal, er, mettle to Queen Victoria.) That aside, do you have to take it off when you go through an airport detector? And, more to the point, what's it for?

"It doesn't," smiled the Goya of the glans, the Leonardo of the labia, the, er, Closterman of the clitoral hood, "take much imagination to imagine what you can do with a good solid piece of gold or stainless steel attached to the end of your, ah, manhood."

Not as much as you might think, apparently. You can't use it as a remote car-lock or to pick up Radio 2. But you can put a chain on it and be led around the room by your, ah, manhood. The piercer's, ah, manhood was one of the few anatomical terms I understood in a sequence that was otherwise awash in inner and outer labia and miscellaneous urethrae. But the combination of technical terms and hearty euphemisms aptly conveyed the spirit of erotic activities which were active but strikingly non-erotic.

Ultimately, I suppose, that's a matter of personal taste (and anyone who disagrees is welcome to come round and walk across my back in stiletto heels while I'm wearing my Prince Albert and peep-hole bra). But, for all the criticism of *Henry And June*, the new film at least understands that the essence of eroticism is adventure. The S&M and infantilism in "Sex Talk" seemed so burdened by rules and conventions that there's no room for adventure: when you want the whipping to stop, for example, instead of shouting "Stop! Ooh! Aaow!" (which means you're enjoying it) you give an agreed code expression like "Douglas Hurd". But, to some of us purists, that bears the same relationship to a real sexual experience as a Territorial Army exercise does to the coming Saddamite cataclysm. Come to think of it, the same could be said of programmes like "Sex Talk". Dissecting it in public removes the shame but also a lot of the pleasure.

BALLS

Mordecai Richler

July 5th 2001/September 2001
The National Post/ The New Criterion

I TREMBLE to say anything nice about Mordecai Richler, since one of my favourite columns of his was his merciless round-up last fall of selected gems on the late Father Of Our Country: "Now Pierre Elliott Trudeau has embarked on his last canoe trip into the wilderness. He is gliding down the dark waters of the river Styx, searching for his lost child. They will soon be reunited." Now Mordecai has embarked on his last pub crawl. He is careering erratically down the dark waters of the river Styx, searching for Nick Auf Der Maur and a decent waterfront bar.

I'm glad *The National Post* reprinted Noah Richler's wonderful memoir of his father from a 1992 British anthology. But two things should be added: first, aside from Noah, pretty much everything else in the book was unreadable, beginning with the awkward title, *Fatherhood: Men Write About Fathering*, and continuing through the forced New Man empathy-flaunting by an interchangeable North London literary coterie. Noah was much jeered at by metropolitan cynics for his childhood memories of joint visits to the bathroom – "I remember the shadow of his enormous dick to the left of my eye" - but, speaking as a father, I can think of few things I'd rather have my son write about me. Except, of course, for Noah's next line: "I love my Pa." He wrote about his Pa affectionately and unaffectedly.

The second thing to add is that I wish *The National Post's* headline writer had stuck with the original title of the essay: "His Balls."

If ever there's a phrase that sums him up, that's it. Only Noah and his siblings were in the shadow of his enormous dick, but we were all in the shadow of his balls. For one thing, he was, as the American critic James Wolcott said, a "horny" writer, the horniness palpable from the earliest work right up to *Barney's Version*. A novelist has to understand sex, for it's the launch pad for so many of our pathetic hustles and deceptions. We once discussed Hugh MacLennan, the pre-eminent Anglo Quebec novelist of Mordecai's youth. "Hugh was such a decent man," he said, "but such a boring novelist. The sexiest he got was in *The Watch That Ends The Night*, but even there the heroine had breasts like inverted champagne glasses." Not Richler's.

But Richler had metaphorical balls, too. He was not in any way a conservative, but he had no time for the faintheartedness of a liberalism so defensive that, as he wrote in 1959, it couldn't bear to contemplate "a Negro whoremonger, a contented adulterer, or a Jew who cheats on his income tax, buys a Jag with his ill-gotten gains, and is all the happier for it." The straitjackets of identity politics have only tightened in the four decades since he wrote those words. Had he not been Mordecai Richler, I doubt whether he could have snuck certain passages of his recent work past the censorious pen of the average Canadian editor.

He was least correct of all when it came to his own country, despite the stampede from the grand panjandrums of "CanLit" to conscript him posthumously to the ranks of "Canadian novelists", Mordecai was a novelist who happened to be Canadian, which isn't quite the same thing. He spent much of his life making gleeful digs about all the great writers who were, as he put it, "world famous all over Canada". Richler, by contrast, was world famous in, among other places, Italy, where his last novel, *Barney's Version*, is a bestseller in its seventh printing and hugely popular among a population not known as great novel-readers. The word "Richleriano" has become the accepted shorthand for "politically incorrect".

****

Richler was certainly Richleriano. In *Solomon Gursky Was Here*, there's a scene set in the early Seventies in which one middle-aged character, forced to play host to a gay son and his lover, staggers drunk into the bathroom to check the pencil mark he's drawn on the jar of Vaseline. His wife is broken-hearted, he's filled with disgust. "It's not that I'm prejudiced against faggots, it's just that I don't like them," he says, pouring himself another Scotch. It is a satirical moment, but the pain underpinning it is true in a way that the standard approved supportive bland uplift is not. Yet I wouldn't bet on any tyro novelist's chances of getting that scene past a North American editor, most of whom are decidedly non-Richleriano. Pat Carney, a Canadian Senator and former cabinet minister, wrote a memoir last year and discovered after publication that a change had been made to the passage detailing her father's job on a merchant ship to China: the editor had reflexively amended "fireman" to "firefighter". The Senator's father was not, as the ensuing paragraph made clear, a man who drives a municipal fire truck and squirts a host at blazing buildings, but a man who stokes the ship's furnaces with fire-making coal. Nonetheless, better to get things wrong than risk even the most hypothetical offence. A passing remark that the drab dress Canada's Governor-General wore for the Speech from the Throne made her look like a washerwoman was struck down, over the author's protests, on the grounds that "it suggests all Chinese-Canadians worked in Chinese laundries". (Her Excellency the Viceroy was born in Hong Kong.)

Against this grubby world of feeble evasions and genteel absurdities, Mordecai Richler stood firm. What's impressive is how much of this foolishness he foresaw so long ago, almost half a century now, in the novel in which he found his voice, *The Apprenticeship Of Duddy Kravitz* (1959). Two decades later, Kerry McSweeney pronounced magisterially in the journal *Studies In Canadian Literature* – what, you don't subscribe? – that "however memorable, *Duddy Kravitz* is hardly a masterpiece." Compared to what? As the years go by, *Duddy* seems more and more remarkable in its anticipation of contemporary fads. Take Virgil, the novel's kindly epileptic, who wants to be proud of his condition and to that end starts a magazine for epileptics with features such as "Famous Health Handicappers Through History":

> *No 2: A Biography Of Julius Caesar... Life was no breeze for the young Julius, but from the day of his birth until the day he met his untimely end he never once let his health handicap stand in his way. Julius had been born an epileptic and was not ashamed of it. He had guts-a-plenty.*

This is 1959: decades before the ADA, before "differently abled" and "visually impaired", before the FDR memorial in Washington got bogged down by accusations that they weren't placing sufficient emphasis on his polio...

The Richleriano aspects of his career fall into three phases: first, he offended Jews; then, English Canadian nationalists; finally, Quebec separatists. In the early Jew-offending phase, he gave a lecture to a Jewish audience at which someone asked: "Why is it that everybody loved Sholem Aleichem, but we all hate you?" The answer to that hardly needs explaining. Mordecai remembered every detail of his working-class childhood in Montreal and the rare glimpses he got of the would-be gentrified Jews in the suburbs of Outremont and Westmount, and some of those details were too funny to let go. In *Solomon Gursky*, it all comes together at the 75th birthday gala of the Bronfmanesque Bernard Gursky, a coruscating dissection of the charity banquet circuit –

> *...plaques, plaques and more plaques, which they awarded one another at testimonial dinners once, sometimes twice a month... They took turns declaring each other governors of universities in Haifa or Jerusalem or Man of the Year for State of Israel Bonds. Their worthiness certified by hiring an after-dinner speaker to flatter them for a ten-thousand-dollar fee, the speaker coming out of New York, New York; either a former secretary of state, a TV star whose series hadn't been renewed or a Senator in need.*

Bernard Gursky's birthday is the *ne plus ultra* of such events: the medley from Jimmy Durante; the Israeli Ambassador's presentation of a Bible encased in hammered gold, the flyleaf signed by Golda; his wife Libby's rendition of "their song", which is inevitably "Bei Mir Bist Du Schön", with the "Bella, bella" line changed to "Bernie, Bernie".

An official of the Canadian Football League passed Mr Bernard a ball, a memento of last year's Grey Cup game, that had been autographed by all the players on the winning team, and then one of the team's most celebrated players, a behemoth who peddled Crofter's Best in the off-season, wheeled a paraplegic child to the head table. Mr Bernard, visibly moved, presented the ball to the boy as well as a cheque for five hundred thousand dollars. Three hundred guests leaped to their feet and cheered. The boy, his speech rehearsed for days, began to jerk and twist, spittle flying from him. He gulped and began again, unavailingly. As he started in on a third attempt to speak, Mr Bernard cut him off with an avuncular smile. "Who needs another speech," he said. "It's what's in your heart that counts with me, little fellow." And sotto voce, he told the player, "Wheel him out of here, for Christ's sake."

Canadian Jews didn't care for such stuff, and called the Canadian Jewish Congress to see what could be done about Richler. The CJC sent him a note, but he didn't take too much of it in, being too distracted by the letterhead: "Cable address JEWCON."

Richler liked to say he emerged from two ghettoes – one Jewish, one Canadian – or to put it another way: one highly marketable, one of little interest. I would rank him above Philip Roth et al, if only because the Canadian qualification of his Jewishness gave him an insight into the points where identities intersect, where the perspective shifts. One of my favourite Richler characters is Mortimer Griffin, the protagonist of *Cocksure*. Mortimer is a middle-class Anglican from the town of Caribou, Ontario, who's made it big in swingin' London as the editor of the small but influential Oriole Press.

But, when the firm gets bought up by a Hollywood mogul, Mortimer suddenly finds his life freighted by Jewishness. On the one hand, for the first time in his life, he's the odd one out, because he's not Jewish, which was never a problem back in Caribou. On the other hand, everyone seems to assume he is. A man from *Jewish Thought* starts following him around accusing him of being a self-hating Jew who's swapped his real name for something more Anglo; his London friends corroborate the story by pointing out how anti-Semitic he is; his black secretary tells him she won't sleep with Jews and he can't prove he isn't one because he's circumcised. To make matters worse, his wife is cheating on him with a man she thinks is Jewish, but is actually a bloke called Gerald Spencer who figured it would be a good career move to change his name to Ziggy.

Richler wrote *Cocksure* at the close of a 20-year sojourn in London. He moved back to Montreal to find himself in the midst of an alleged Canadian Renaissance – a cultural flowering of a young nation eager to cast off both colonial ties to the Mother Country and the cultural oppression from the south. It's very difficult typing that sentence without tittering. Be that as it may, while Richler was out of the country, Canada had acquired the Governor-General's Awards for Literature, Canada Council grants for writers, and an elaborate public subsidy racket for Canadian publishing houses that ensures to this day that no matter how bad a book is a publisher has very little incentive not to publish it.

Things were different when young Mordecai was shopping his first effort:

> *Then, Andre Deutsch, Ltd, the British publishers, made an offer for my novel. A conditional one, however. They would publish* The Acrobats *if I agreed to do more work on it. I was offered an advance of £100 (approximately $275) - £50 on signature of contract and another 50 once my revision had been found acceptable. I sent an immediate cable of acceptance.*
>
> *"I don't get you," my Uncle Jake said. "You put two years into*

writing a book and now you're happy because some jerk in London has offered you a lousy two-fifty for it. You could have earned more than that cutting my lawn..."

Richler was dispatched by Andre Deutsch to Toronto, to impress the Canadian distributor, who instead pointed out the grim truth:

No serious Canadian novelist - including Morley Callaghan or Hugh MacLennan - is able to support himself strictly on the sale of his novels in Canada. The distributor was prepared to risk a first order of 400 copies for all of Canada. I stood to earn approximately $32, if they sold out.

All the above was true, but already irrelevant. Richler, whether consciously or not, was already writing for the world, his publishers, editors and agents in London and New York pre-dating those he eventually found in Toronto: in the Fifties and Sixties, he wrote about Canada from London, distance bringing his experience into focus, sifting and sorting. In any case, he didn't write about "Canada" – whatever that is – only the particular and isolated corner of the Jewish quarter of anglo Montreal in francophone Quebec in the Canadian section of North America; a ghetto in a ghetto in a ghetto in a ghetto. St Urbain Street is as foreign to any Torontonian or Winnipegger as it is to a New Yorker, Dubliner or Parisian. In the Ivy in London or the Polo Lounge in LA, Richler and his fellow scribblers (his preferred term) could talk about novels, deals and money, but, when he showed up in Vancouver or Edmonton people just wanted to badger him about (dread phrase) Canadian identity. "Special pleading," he sighed, "whether by Canadian sports writers, kibbutzniks in Galilee, or proliferating Canada culture boosters, never fails to move me to mockery."

On the off-chance that you haven't read Hugh MacLennan's essay *The Psychology Of Canadian Nationalism* - a title so flamboyantly off-putting as to raise suspicions that he was deliberately trying to reduce his royalties as some sort of tax wheeze – let me explain that the author's principal point is that a "feminine psychology" runs through

Canadian literature: the Americans are the masculine in North America, Canadians the feminine. Most of us can see where he's going, and that's all we need to know, thank you very much. Nonetheless, Michael Valpy in Toronto's *Globe And Mail* felt obliged to expand the thesis the other day. He notes that America's most famous adolescent is Huck Finn, a boy bustin' to "light out for the territory", beyond the constraints of the civilized world, while Canada's most famous adolescent is Anne of Green Gables, a girl so domesticated that her address is part of her name. According to Valpy, the defining emblem in American culture is the horse, in Canadian culture the house: "To be on a horse is to move. To be in a house is to be fixed." Actually, a horse can be fixed too, but that's another story. "It is an American, Thomas Wolfe, who says you can't go home again," Valpy continues. "It is a Canadian, Morag in Margaret Laurence's *The Diviners*, who says you have to go home again to be in harmony with life."

As a column intended to give you one more reason not to read Canadian masterpieces, Valpy's is a masterpiece itself. Unfortunately, it seems he intended precisely the opposite effect, blithely assuming that his house/horse hockey would make readers chuck their Wolfes and rush out to buy *The Diviners*, if only to check whether Morag expresses herself quite as clunkily as Valpy claims. If it is true that Canadian literature is feminine and house-bound, then Mordecai Richler was not a terribly Canadian writer. He got out of the house. His film credits include *Room At The Top*, a quintessentially northern English kitchen-sinker with Laurence Harvey from those gloomy, monochrome, petrol-rationed British Fifties, and *Fun With Dick And Jane*, a quintessentially garish Hollywood comedy with Jane Fonda. In his novels, Mordecai was a kind of Huck of Green Gables, secure enough in his sense of home to "light out for the territory", the territory in this case being London, Paris, New York, Hollywood, Spain, and beyond - because for a Canadian, unlike an American or an Englishman, if you don't go to the world, the world sure as hell isn't going to come to you: that's why Canucks, not Yanks, are the great mid-Atlanticists – the high yallers, the octoroons of the world's great cities, able to pass wherever they go.

Richler's life skewers the delusions of the Valpy school. If there is a Canuck Huck, it's Duddy Kravitz. And, if Canadians cling to home and security, what are we to make of *Solomon Gursky Is Here*? Richler took the Wandering Jew and plunked him in the frozen north, placing one of the chosen on Sir John Franklin's lost expedition to the Northwest Passage in 1845. He'd been reading a lot of "magical realism" and called *Gursky* "the first South American North American novel": it's magical realism with jokes, a surprisingly effective combination and perhaps the closest to the Great Canadian Novel we'll ever get. It does everything Valpy says Canadian literature doesn't do, wandering off to London, Washington, Entebbe, the Polar Seas, as, indeed, Canadians do.

But in the 24 hours after Mordecai's death the grand subsidy-fatted bores of Canadian letters lined up to insist that, in fact, the guy was very much a "Canadian writer". Even his Canadian agent, Louise Dennys, lapsed into the usual shtick:

> *Mordecai mentioned that Canada was to him a fragmentary place in the Seventies. I think it's because of Mordecai that Canada is no longer a fragmentary place, not only to ourselves but to other people outside of Canada. He's given it a particular and distinctive shape that we can only be truly grateful for.*

Even if this were true, which it isn't, it would be the least of his achievements. On the matter of Canada's distinctive shape, he and I were once on a radio show in New York with Garrison Keillor, who opined en passant that the reason Americans found it hard to get interested in Canada was that the country was impossible to visualise: it has no recognisable shape. The southern border is just about clear, but the rest of it bleeds away into Queen Maud's Gulf and the Arctic Circle and God knows where. In his last years, Mordecai himself took on the same indeterminate, shapeless quality as his native land, with everything straggling and dangling and spreading over the map - the floppy mop of grey hair, the jowls, the jacket, the hanging belly. He, like Canada, was in much better shape 30 years ago. On the other hand, he was one of the few genuine laugh-out-loud authors and, at a time when the comic novel is in poor health, Richler's reliability in this respect was good news for readers all over the world who have no interest in the non-fragmentariness or otherwise of Canada. He was rooted in Montreal but, unlike the general tenor of his obituaries, he was not parochial.

He jeered not just at the stunted nationalism of Quebec but at the moral smugness of modern Canada, which was why the state had such difficulty paying tribute to him. The Prime Minister's statement would have had him weeping with laughter: "Mordecai Richler was the quintessential Canadian man of words, and his loss leaves us grasping for words that can do justice to his importance in Canada's artistic landscape," said Jean Chrétien, grasping for words and eventually deciding that the guy "made us all proud to be Canadian." The charitable interpretation is that this is the standard tribute for Canadian "men of words" and that whoever wrote it had never read a word of Richler, but it would be more pleasing to think they'd read it all too well and decided anyway to kiss him off with the usual lame-o CanLit boosterism. As for making us all proud to be Canadian, here's how one Richler character summed the place up:

> Let me put it this way. Canada is not so much a country as a holding tank filled with the disgruntled progeny of defeated peoples. French-Canadians consumed by self pity; the descendants of Scots who fled the Duke of Cumberland; Irish, the famine; and Jews, the Black Hundreds. Then there are the peasants from Ukraine, Poland, Italy and Greece, convenient to grow wheat and dig out the ore and swing the hammers and run the restaurants, but otherwise to be kept in their place. Most of us are huddled tight to the border, looking into the candy store window, scared of the Americans on one side and the bush on the other.

The irony in all this is that, in the end, Mordecai was one of the few writers in the world who can claim to have saved his country. In the Nineties, irritated by Quebec separatism, he started writing about its oppressive triviality: arriving outside a pub in Montreal one day, he found an agent of the Office de la Langue Française photographing the menu blackboard and measuring the inscription "Today's Special: Ploughman's Lunch". Under Quebec law, signs can only use English words if they're half the size or less of the accompanying French words. An essay for *The New Yorker*, later expanded into a book, *Oh, Canada! Oh, Quebec!*, caused particular distress to the Parti Québécois, who never forgave Richler for, as they saw it, making them a laughingstock in the outside world. In fact, the outside world never gave Quebec a thought. No doubt in Manhattan there were those who marveled, "Can you believe it? There's a long piece in *The New Yorker* this week that's actually readable!" Then they promptly forgot about Quebec.

Back home, though, where Anglophones had reacted to separatists either by enduring their humiliations (the so-called "lamb lobby") or by fleeing to Toronto, Richler's essay legitimized scorn, while the PQ's outraged reaction to the puncturing of their prestige only emphasized the pitiful state of the nationalist movement. In the 1995 referendum on secession, the final result was separatists 49.5%, federalists 50.5%. It's not too fanciful to assert, in a tight poll, Richler made just enough difference to save the day. I happen to disagree with him on separatism, since on balance I find smug English Canada the more insufferable, but even so Quebec boasts the world's dumbest secessionist movement, forever trying to explain to its citizens why we need to set up our own country exactly the same as the one we'll be leaving. But there's no doubt Richler the "controversialist", the "misanthrope", the "curmudgeon" did more for Canada than all the sunny maple-draped multiculti CanLit boosters put together.

At his home in the Eastern Townships, he'd repair to the Owl's Nest every afternoon and enjoy a little light banter with the bar's cheerfully unreconstructed clientele – plumbers, carpenters, leathery truck-driving women. "Are they better company than Martin Amis?" I asked him.

"I wouldn't say that," he said. "They're not as witty. But they're more interesting, it's a richer world. The literary world, in the larger sense, is dull: it's a tradesmen's world." He came home not to be a "Canadian writer," but to be a writer. The Montreal he wrote about is fading fast. As he pointed out, he's far from the only Anglophone Jew whose kids have all fled to Toronto, London, wherever. What matters is not that he gave shape to Canada, but that he gave shape to his own fictional landscape. One day, soon, St Urbain Street will be as lost to the real world as P G Wodehouse's Drones Club or Blandings Castle. But, like Wodehouse's, it will endure.

In his column about Trudeau's passing he bemoaned the "national insecurity" that required "outsiders to confirm that Canada exists," mocking *The National Post*'s box score on how dead Canadians were treated by *The New York Times*: Trudeau was the first to make the front page since Glenn Gould in 1982. Diefenbaker was relegated to page 16, Robertson Davies to page 9 of the second section. The day after his own death, there was Mordecai on page one: the "quintessential Canadian man of words," but not so quintessentially Canadian that Americans, English, Irish and many others won't miss him, too.

LEG

Breaker

July 15th 1989
The Independent

"**N**O STATUE has ever been put up to a critic," Sibelius is supposed to have said, and to every slighted artist it affords some consolation, even though it rather overlooks how few statues there are to writers, composers and painters, at least when compared to kings, generals and prime ministers: you wouldn't, in other circumstances, find artsy-fartsy types so eager to endorse the values of public statuary.

In any case, not all critics want for honours: Brooks Atkinson, the *New York Times* man from 1924 on, has a theatre named after him; his predecessor Alexander Woollcott is commemorated by a cocktail, the Brandy Alexander. Few of us loitering in their footsteps are likely to be celebrated in either fashion (obligatory false modesty). But it is a diverting game, ideal for whiling away the longeurs in Tommy Steele's production of *Singin' In The Rain*, to divide today's generation of drama critics into those whose immortality would be more aptly conferred by the theatre marquee or the stalls bar.

Woollcott, though, has an additional honour denied to Atkinson. "He was a weekend guest at Moss Hart's place," says the director Gene Saks, "and he got very pissed off that Max Gordon, the producer, was there, sharing the billing as star guest. Woollcott ordered the cook to serve him in his room and even invited other guests to have dinner up there." On departure, he wrote in Hart's guest book, "This is to certify that I had one of the most unpleasant times I have ever spent." Recounting the tale to George S Kaufman, Hart said,

"Wouldn't it be awful if he'd broken a leg and had to stay the summer?"

The electric light bulbs flashed in their brains simultaneously and the two got to work.

In *The Man Who Came To Dinner*, Woollcott becomes Sheridan Whiteside, laid up in small-town Ohio, at the home of Mr and Mrs Stanley. He takes over the house, abuses his hosts and makes them eat upstairs in the bedroom. Woollcott, flattered by this characterisation, demanded to play the role on tour.

"We'd all like to have the nerve he has," says Saks, who's staging the Royal Shakespeare Company's revival of the play, "to put someone in his place with a scathing insult, and to tell off all the Mr and Mrs Stanleys of the world, these stupid, proper, bourgeois burghers."

Four years later, in 1943, Woollcott's love of Brandy Alexanders, rich food and a generally hedonistic lifestyle contributed to his fatal heart attack at 56. Brooks Atkinson, seven years younger, lived on until 1984, by which time his profession had vastly increased in importance - or, at any rate, more effectively communicated its self-importance. At dinner recently, an earnest RSC actor asked me, "Would it help you and your fellow critics if you came to rehearsals and worked through the play with us?" Appalled at the prospect but unable to produce any sufficiently serious excuse, I slid under the table and passed out.

Today, everyone takes drama criticism so seriously. Seventy years ago, no-one minded *Vanity Fair*'s critic, P G Wodehouse, reviewing shows by that up-and-coming dramatist, P G Wodehouse. Nor did they object to Woollcott's habit of loading his reviews with brand-names and accepting gratuities for the favour. Harold Ross, editor of *The New Yorker*, once intercepted 100 neckties sent in return for a lavish plug and distributed them to his staff to wear for Woollcott's next appearance at the office.

Woollcott had met Ross in Paris during the Great War, when both men had been conscripted on to the editorial board of the morale-boosting *Stars And Stripes*. It was a demanding job, although not journalistically. Woollcott, Ross and John T Winterich were ordered by their commanding officer and editor to interview a visiting Congressman for the paper. "Why three of us?" asked Woollcott.

"He's a pacifist," said the CO. "After you've interviewed him, beat him up."

Naturally enough, the aesthetes proved unable to discharge this task. An effeminate man, Woollcott spent much of his life reading sly accusations of homosexuality into the multitude of uncomplicated insults fired at him: "Louisa M Woollcott," the lyricist Howard Dietz dubbed him. Those who derive pleasure from the old gag that critics are the eunuchs in the harem – they're there every night, they know how it's done but they can't do it themselves - may be interested to note that Woollcott was, it seems, impotent.

E B White described accompanying the critic, silk hatted, cape billowing, to a first night as equivalent to "being towed in a dinghy behind a large and very expensive yacht". Today, his successor, Frank Rich, maintains a saintly detachment from Manhattan's glittering theatre set, and Broadway first nights are critic-free zones: reviewers attend on previews, because they now, apparently, require some days to hone their judgments to precision. Woollcott, in contrast, would despatch the play to oblivion before the celebrities invited back to his apartment had got through their first drink. "He finished his review in nothing flat," wrote Howard Dietz of one occasion, "and sent Junior, his worshipful mulatto, to the drama desk of the *Times*."

The lifestyle is more impressive than the criticism. Woollcott belonged to that generation of Algonquin wags still treasured for their devastating dismissals. Dorothy Parker: "*House Beautiful* is play lousy", etc. But his own offerings are not quite in the same league, and, for the darling of ladies' luncheon clubs, strikingly single-minded: "*Number Seven* opened last night," he wrote, and, alluding to an American euphemism for defecation, continued, "It was misnamed by five"; *And I Shall Make Music* was booted into touch in one line: "Not on my carpet, lady."

Otherwise, his prose is as florid as Kaufman's is lean. *The Man Who Came To Dinner* captures the style perfectly:

"One misty St Valentine's Eve," begins Sheridan Whiteside, nostalgically, "- the year was 1901 - a little old lady who had given her name to an era, Victoria, lay dying in Windsor Castle. Maude Adams had not yet caused every young heart to swell as she tripped across the stage as Peter Pan; Irving Berlin had not yet written the first note of a ragtime rigadoon that was to set the nation's feet a-tapping…"

Well, you get the idea: the largely irrelevant achievements of his pals, preceded by the author's favourite all-purpose historical reference. "In England," the real Woollcott begins, "just before the turn of the century the old Queen was resting from the Diamond Jubilee, whereby, with agreeable but exhausting affection, her subjects had just celebrated her 60 years upon the throne…" - and all this just to beef up some gush on A E Houseman. He did come up with a cute handle for Clifton Webb, who specialised in ballads of resignation – "the general futility man". But who now knows or cares about Clifton Webb?

And what of Woollcott? A new encyclopaedia of theatre includes him but spells his name wrong, which would have infuriated him. Indeed, the celebrated Algonquin Round Table only began as a gag against Woolcott in which a luncheon invitation was sent out, listing a series of wartime reminiscences by Alexander Woollcot, Woolcott, Wolcott, etc.

It was an environment in which a smart Aleck (the phrase dates from the 1860s, but Woollcott reactivated it) could thrive. He had no taste (he thought Eugene O'Neill was drivel), but he was the highest paid critic of his day; he had a squealing voice, but was the highest paid radio commentator; he boasted that his columns on famous murder cases had made him more money than the lawyers.

"He doesn't deserve to be remembered for his writing," says Gene Saks. "But he's one of the most outrageous characters we've ever had. And I think men also deserve to be immortalised for their chutzpah."

"Life without Aleck," Kaufman told Hart after the great man's death, "is like a play with a crucial scene dropped. It still plays, but something good has gone out of it."

THIGHS

Return to disaster

July 25th 1996
The Daily Telegraph

IF ANY SCENE sums up the disaster-movie genre it's Shelley Winters swimming underwater through a flooded corridor in *The Poseidon Adventure*, her cheeks puffed out like a blowfish, dress billowing up over flailing thighs. *Newsweek* ungallantly observed that she's "plump enough these days to sink an ocean liner all by herself", but Miss Winters insisted that "I put on all this weight for the movie!" and her deal required the studio to pay for post-shooting sessions at a fat farm. If they did, they deserved a refund. Shelley stayed plus-sized and (just when you thought it was safe to go back in the water) resurfaced in *Tentacles*, in which she got the better of a giant squid.

Unlike Shelley, the disaster movie itself shrivelled away to nothing. It was the only new film genre to emerge from the 1970s, and, like everything else from that decade - Abba, Edward Heath – it's been dusted off for the Nineties. The original disaster movie is usually reckoned to be *The Poseidon Adventure* (1972), though you could make a case that the prototype was *San Francisco* (1936), in which Jeanette MacDonald's singing sets off the 1906 'Frisco quake.

At any rate, Jan De Bont's attempted revival of the genre, *Twister*, preserves most of the essentials of the form, albeit without Shelley's flailing thighs and with one or two variations. In these films, man's ambition is derailed by an appalling natural disaster - a ferocious tornado, a tidal wave, an erupting volcano, O J Simpson's acting. A propos the latter, unlikely as it seems, terrified people once placed 911 emergency calls and hoped O J would turn up in time - as opposed to these days, when terrified people call 911 because O J has turned up.

But back in 1974, in *The Towering Inferno*, O J was the fellow you cast if you wanted someone to play a security guard who rescues two children and a cat from a blazing bedroom.

It's always two kids. One is never enough. Shelley Winters established the tradition in *The Poseidon Adventure*, and, by the time Burgess Meredith carried a brace of moppets across a burning bridge in *When Time Ran Out*, it was de rigueur. By this stage, in defiance of the old showbiz saw about children and animals, most big-time movie stars were insisting on contractual guarantees that they be allowed to rescue two infants and a family pet. Invariably, going back for the dog proves the hero's undoing. In *Twister*'s opening sequence, it's the pooch that does for dad.

That's the funny thing about this multi-million-dollar blockbuster. It spends a fortune re-creating all the clichés we took for granted 20 years ago. For example, *Twister* has the bit where the telegraph lines come down and electric wires dance in lethal convulsions across the highway - just like *Earthquake* (1974). There's also the moment when some humdrum everyday item becomes technologically vital. In *Twister*, it's Coke cans - which, frankly, is a bit tame. In *Earthquake*, Lorne Greene, in the midst of the rubble that has buried every green lawn, barks at his secretary: "Take off your pantyhose, dammit!" And, with her nylons providing the final crucial link in an elaborate pulley system, he ferries everyone to safety. Well, not everyone. Traditional examples of the genre are like a sort of all-star balloon debate, in which the best a supporting actor can hope for is to hang in there a little longer than his billing merits.

Irwin Allen, father of the disaster movie, made his first all-star grab-bag in 1957. In *The Story Of Mankind*, Hollywood's biggest names played history's most fascinating figures: Hedy Lamarr as Joan of Arc, Harpo Marx as Sir Isaac Newton, Dennis Hopper as Napoleon. By the Seventies, he'd figured out that there was far more money to be made getting Hollywood's biggest names to play cardboard characters of no interest whatsoever.

As far as casting is concerned, the trick is to match the star with the most unsuitable occupation: Dean Martin as a pilot, Charlotte Rampling as a marine biologist, Jacqueline Bisset as anything. And if you're wondering what all these people do in a volcanic eruption or a dam burst, that's easy: they work out their personal problems. Nothing like dodging molten lava to fix up your marriage or cure you of substance abuse.

Twister doesn't mess with this convention: at the start of the movie, Helen Hunt and Bill Paxton are on the brink of divorce; luckily, a huge tornado blows in and, like a jollier Claire Rayner, brings them to their senses. You'll notice that *Twister* has dispensed with the all-star format: Helen Hunt is a pleasant enough actress from a nondescript sitcom; Bill Paxton is one of those actors you never remember, or, if you do, it's because you're mixing him up with Bill Pullman (the one in *Independence Day*).

As the presence of Shelley Winters in *Tentacles* confirmed, the Seventies disaster movie spawned an even more lucrative sub-genre in underwater disasters – deep-sea epics of fish with chips on their shoulders. In the wake of *Jaws* came *Barracuda*, *Tintorera*, *Orca* and *The Deep*. Eighties environmentalism left the fish pic beached by the tides of political correctness - so we wound up with *Free Willy*, in which the killer whale is the good guy and the plot revolves around getting him back in the water.

Fish aside, the problem with the genre was that there was only room for one movie per disaster. Irwin Allen recognised this in 1974. While he was adapting a book called *The Glass Inferno*, he learned that Warners had bought the rights to a similar book called *The Tower*. Cannily, he suggested a merger: *The Towering Inferno*. *Variety* suggested they go a stage further, combining *The Towering Inferno* with *Earthquake* and calling it *Shake'n'Bake*. Disaster movies spluttered on until *When Time Ran Out* (1980), by which time it had.

Typically, the British didn't notice, and it fell to Lew Grade to make the most spectacular disaster movie ever: *Raise The Titanic!* "It would have been cheaper if we had," said Lord Grade afterwards. It lost

$29.2 million, which would have made it the all-time box-office disaster. But Grade's film was so disastrous that even that distinction eluded it: the same year, 1980, *Heaven's Gate* came along and blew $34.2 million.

If you want proof that Hollywood has recovered its sea legs since Lew Grade went down with all hands, look no further than James Cameron's next big-budget epic: *Titanic*. He's got Kate Winslet (*Sense And Sensibility*) and Leonardo DiCaprio (from *Romeo And Juliet*), but it's still early days. Shelley Winters as the iceberg?

KNEE

Pads

September 26th 1998
The Spectator

FTER THE third hour or so, it starts to seem perfectly normal to see the President of the United States with a "Sexually Graphic Material" network warning slapped across his chest as he attempts to explain what sex isn't:

"If the deponent is the person who has oral sex performed on him, then the contact is not with anything on that list, but with the lips of another person," asserted Bill Clinton. "It seems to me self-evident that that's what it is. Let me remind you, sir, I read this carefully."

Like Groucho and Chico, Clinto has fine-combed the small print: if the party of the first part is apart from the parts of the party of the second part while the party of the second part is partaking of the parts of the party of the first part, then even though the party of the second part is taking part with a particle of the party of the first part, unless the party of the second part is partly parturient, the party of the first part plays no part - though whether this will fly with the party of the first part's party (the Democrats) is another matter.

By the fourth hour, I was bored stiff: "Let's go into the anteroom," I said to my secretary, "and play Deponent and Intern."

We couldn't find the word "deponent" anywhere in the *Kama Sutra* - and we've got the original edition, not the revised *Clinta Sutra*, which is a slim pamphlet containing a brief description of the missionary position; on the other hand, its companion volume, *The Joy Of Non-Sex*, goes on for thousands of pages.

Unfortunately, in the Clinton video, as with most sexually graphic material, to get to the hot stuff you have to sit through a ton of perfunctory, implausible dialogue:

> *I don't recall... I have no recollection whatsoever of that... I don't remember anything I said... I'm not saying I didn't, but I have no recollection... I have no specific memory... I remember specifically I have specific recollection of two times, but I don't remember when they were... I've told you what I remember. It doesn't mean that my memory is accurate... I do not remember when they were or at what time of day they were or what the facts were... I do not know what I meant...*

Understandably perhaps, he found it hard keeping track of his women: at one point, he referred to the Jones deposition as "the Lewinsky deposition"; at another, "the Flowers deposition". So many gals, but only one hard-pressed deponent. "I have been blessed and advantaged in my life with a good memory. Now, I have been shocked, and so have members of my family and friends of mine, at how many things that I have forgotten," the President revealed. "Compounded by the pressure of your four-year inquiry, and all the other things that have happened, I'm amazed there are lots of times when I literally can't remember last week." In other words, I'd be able to remember everything you guys wanted to ask me about if only you'd quit asking me about it.

For the benefit of the President, what happened last week was this: Senate Minority Leader Tom Daschle and other senior Democrats told him he had to cut out the evasions and legal hair-splitting; it was an insult to the public. But on Monday, the public saw four hours of evasions and legal hair-splitting and what happened? His approval rating bounced from 59 per cent to 68 per cent, and even his personal approval rating went up from 37 per cent to 44 per cent - the first halt to the steady erosion in his poll numbers since the mea sorta culpa of August 17th. The wise old Watergate hands point out that public opinion coalesces slowly - which is more or less what's been happening over the last month. But now, even this painfully slow trickle away from the perennial Comeback Kid has been reversed. Why?

Well, for one thing, although recorded only a few hours before that disastrous address to the nation, his grand jury testimony was very different: he was more contrite, he actually uttered the words "I'm sorry", and he expressed sympathy for "Monica". "She's basically a good girl," he said, and, though mindful of feminist sensitivities he quickly corrected himself to "good young woman", the original formulation humanised him in a way that his "National Prayer Breakfast" and signing Jesse Jackson as his family's spiritual adviser failed to. Granted that he's a shifty lyin' four-flushin' crock-peddlin' jive-ass sonofabitch, had he adopted the tone of his grand jury testimony in his subsequent speech to the nation, he would have spared himself this last wretched month.

Alas, the other reason for the President's revived numbers is more prosaic: his spinners are spinning again. They were thrown from the runaway horse for a while, but they're back in the saddle now. Last weekend, the consensus was that the White House had been outspun. If only they'd agreed to let the President go to the grand jury in person, there'd have been no videotape for the "partisan" Republicans to release in the first place. "Authoritative sources" who'd seen the recording assured the networks that Mr Clinton was profane, he lost his temper, he stormed out of the room, but on Monday morning when the tape rolled President Godzilla was nowhere in sight. Indeed, it was obvious from Mr Clinton's performance that he knew the video testimony was likely to leak out and was at pains to behave himself.

In spinning the media buzzmeisters into buzzing that the spinmeisters had been outspun, the spinmeisters were in fact playing a cunning game of double-spin: the White House was only pretending they didn't want the video released in order to tar the Republicans as vicious and unfair and to lower expectations of their boy's performance. By Tuesday, when the President's numbers had perked up, the Washington air was thick with rumours of even more elaborate double-bluffs. On Capitol Hill, Mr Clinton's colleagues were suddenly fearful that the White House was planning to leak sex stories about Congressional Democrats to make it look as if the Republicans were playing dirty; in yet another example of the President's amazing ability to corner every position on any issue, the White House is now running dirty tricks for both sides. Mr Clinton is no longer Houdini, but he's doing a passable Rasputin: he's been shot, poisoned, throttled, held under water till the bubbles cease, but he keeps bobbing back up again.

The question is: where can it get him? By Tuesday, previously wobbly Democrats were talking about letting the President cop a plea for "Censure Plus": in exchange for confessing his perjury, the President would get away with censure, plus a fine for the $4.4 million cost of the Monica investigation and/or the cancellation of his pension and/or a misdemeanour conviction and/or loss of his parking space, no interns under 73, etc. In return, the President would get to complete his term. The only trouble with this "compromise" is that no senior Republican is in the least bit interested. Henry Hyde, chairman of the House Judiciary Committee, says such a deal would be a matter for the Senate; having experienced at first hand White House smear tactics with the leaking of a 33- year old affair, he's eager to press on with the impeachment process. The insiders say it's all down to November's elections: if the Democrats lose badly, they'll want to dump Clinton; if it's closer to a tie, Republicans will draw their own conclusions and settle for a deal.

The flaw in this theory is an obvious one: the dynamic in the Clinton scandals is Bill Clinton himself. Only one thing can be said with certainty: something else always turns up. The most artfully drawn deal in the world cannot immunise Mr Clinton against himself - - against the inevitable "second intern" or a new funny-money scandal. The question Americans have to confront is how much longer they're going to subordinate the health of their democracy to the appetites of this President.

After watching the video, Senator Tom Harkin of Iowa became the first Democrat to voice publicly what many of his colleagues have been muttering in private: that the guy has some mysterious "disorder", albeit one not necessarily known to medical science. "I'm coming around to the belief that there's something deeper here, that the President maybe has something wrong with him," Senator Harkin told a television station in Des Moines. "I don't know whether it's an illness… but I don't know how else to explain what he did."

Whatever it is, it seems to be contagious. By any rational measurement, the President's remaining hard-core defenders now sound, in that useful British expression, stark staring bonkers. "Every morning," said the First Lady the other day, "he wakes up worrying about how he can do more to make the American people's lives better." "You have demonstrated, at least in my lifetime, a higher commitment to the kind of moral leadership that I value in public service and public policy than any person I have ever met," said Democratic National Chairman Steve Grossman at a $50,000-per-couple fundraiser just a week ago. "Our prayer for you today," he continued, "is that you will continue to provide the kind of moral leadership to this country that has enriched the life of virtually every citizen."

For the first eight months, this scandal was, as theatricals say, *Hamlet* without the Prince. We were obliged to feast on bit players like William Ginsburg, Monica's goofball lawyer, because the star refused to take the stage. Now, at last, Hamlet stands in the spotlight, cradling Vince Foster's skull in one hand and a Hamlet cigar in the other. As in the play, there is much speculation as to whether he's mad. On the day the Starr report was released, the White House press secretary issued a statement denying that the President was under psychiatric care. Yet increasingly his soliloquies are an assault on sanity: "It depends on what the meaning of the word 'is' is" - or, as Shakespeare put it, "To be or not to be, that is the question."

To Teddy Roosevelt, the presidency was a "bully pulpit": words are its currency. Mr Clinton's post-modern deconstruction of language debauches his currency as surely as a Russian president printing new roubles to replace the worthless old roubles debauches his. The bully in the pulpit lies to the American people, lies to his Cabinet, sends them out to disseminate the lie at taxpayers' expense, and has somehow succeeded in persuading his fellow citizens to place his welfare before that of the Republic: ask not what your country can do for you, ask what your country can do for me. Bill Clinton is a master of words especially where his appetites are concerned. A few months back, asked about his fondness for McDonald's, he insisted that "I haven't eaten at McDonald's a single time since I've been President."

When it was pointed out that, in fact, video footage exists of the President passing through the fabled Golden Arches, the White House explained that the key word was "eaten": though he had certainly been on the premises, it was only for the purposes of beverage consumption.

When it was then pointed out that in 1994 he had declared enthusiastically that "we love to have Egg McMuffins on Sunday mornings", the White House stood by the President's answer but, in a further clarification, explained that the key word was no longer "eaten" but "at": Mr Clinton had eaten McDonald's food, but not at McDonald's; the Egg McMuffins were strictly take-out.

So, although the President has drunk McDonald's beverages at McDonald's and eaten McDonald's food off the premises, his statement that he has not eaten at McDonald's was, by Clinton standards, the truth.

Not surprisingly, for his last deposition, his interrogators were keen to dot and cross every Sausage, Egg and McMuffin:

Definition of Sexual Relations:

For the purposes of this deposition, a person engages in 'sexual relations' when the person knowingly engages in or causes:

(1) contact with the genitalia, anus, groin, breast, inner thigh, or buttocks of any person with an intent to arouse or gratify the sexual desire of any person;

(2) contact between any part of the person's body or an object and the genitals or anus of another person; or

(3) contact between the genitals or anus of the person and any part of another person's body.

'Contact' means intentional touching, etc., etc.

But, for President Houdini, that still leaves plenty of wiggle room, semen on a cocktail dress at the FBI analysis lab notwithstanding.

Speaking of wiggling, the thinking woman's line on Bill Clinton was most pithily encapsulated by Nina Burleigh, a *Time* correspondent who, reflecting on a face-to-face encounter with the Big He, declared, "I'd gladly give him a blowjob to thank him for keeping abortion legal" - thus brilliantly distilling the recent convolutions of feminist thinking. "Have you ever heard of this Nina Burleigh?" I asked my house guest last weekend.

"Yes, she's my best friend," she said, and produced a touching photo of the two of them in the gutted shell of her new house. "Personally, I'd much rather give George Bush a blowjob, but what's he ever done for us?"

This is the way these gals think these days. It seems the President has brought the entire political culture to its knees - the Clinton level. The stain is spreading. At Gettysburg, Lincoln left his mark with a rousing address; Bill Clinton has left his mark by getting aroused on a dress.

In private moments, does he rage like Lady Macbeth? "Out, damned spot! Out, I say!" I doubt it: Mr Clinton is not one for guilt or shame, though he may be increasingly irked at the way he and his spinners have only spun him into a corner. But Bill's in his favourite position now, and it's not sexual. Instead, he's up on stage, all alone, with the spotlight trained on him, ready to look us straight in the camera and tell us his way:

"And now the end is near and as I face the final curtain, my friends, I'll make it clear, I'll state my case of which I'm certain. (I did not have sexual relations with that woman! And, even if I did, oral sex is not adultery, any more than a vanilla milkshake is eating at McDonald's).

KNEEPADS

"For what is a man? What has he got? If not himself, then he has not. It's time to say the things he truly feels [*empathetic Bill*] and not the words of someone who kneels [*Monica*]."

Alas for Bill Clinton, the record shows he took the blows.

CALVES

Polled Herefords

March 27th 1999
The Daily Telegraph

T HE ONLY THING you need to know about Al Gore is that he's the first and only Presidential candidate to rent a herd of cows for the announcement of his candidacy. This was in the 1988 campaign, back in Tennessee, when then Senator Gore decided he needed the cattle as a backdrop for his speech – "We'll want plenty of balloons, bunting, soda pop, recyclable paper cups, some 'Gore '88' buttons... oh, and order up some cows in assorted colours - Holsteins, Jerseys, maybe a couple of Angus..."

"Polled Herefords?"

"Geez, I just want 'em to stand there, I don't need to know their opinions."

Al Gore Jr is the son of Senator Albert Gore Sr. He was born in Washington and grew up living in luxury at the Fairfax Hotel. Young Al's idea of cow pasture was the roof terrace. But a decade after he sent his aides round to Herdz Rent-A-Cow, Vice-President Gore has dusted off his Farmer Al act and taken it on the road. Launching his "Gore '00" campaign in Iowa, he recalled his boyhood for the benefit of *The Des Moines Register*:

> *I'll tell you something else my father taught me. He taught me how to clean out hog waste with a shovel and a hose. He taught me how to clear land with a double-bladed axe. He taught me how to plough a steep hillside with a team of mules. He taught me how to take up hay all day long in the hot sun.*

A year ago, Al Gore was maintaining he was the inspiration for *Love Story*. Now it turns out he was also the inspiration for *Grapes Of Wrath*.

Well, if the most privileged son in American politics wants to pretend he grew up a dirtpoor Tennessee sharecropper eking out a hardscrabble existence, I don't see why we should hold it against him. Readers of this column will know that, in my dispatches from New Hampshire, I'm not averse to pulling the old horny-handed son-of-the-sod routine myself. In truth, I'm only horny-handed because I've been pounding the keyboard all day, tapping out nancy-boy pieces on musical comedy for the *Telegraph* arts pages. But, even if you're going to be a faux farmer, it helps to get the details right. If Al Gore's cleared land with a double-bladed axe, he's the first presidential candidate to do so since Abe Lincoln, known in his youth as the Rail Splitter.

The last time I went out to clear land with a double-bladed axe, I did one sapling, mopped the sweat from my brow, and had to lie down for a couple of weeks. Then I put the double-bladed axe away and took out the chainsaw. When I got tired of that, I called my neighbour and borrowed his brush-hog and stump grinder. A double-bladed axe is fine for clearing the Palm Court of the Fairfax, but out in the country you want something that'll do the job properly. As for ploughing a steep hillside with a team of mules, it's picturesque in a rugged sort of way, but my advice is skip it. Plough anything steeper than a 12 per cent grade and your topsoil's only going to get washed down the hill to your neighbours. Of course, in Al's case, he can just rent it back by the hour as and when he needed it.

Unfortunately for the Vice-President, his touching portrait of a haymaking, mule-ploughing, hog-hosing boyhood came hard on the heels of his claims to have invented the Internet, and to be Courtney Love's biggest fan, and to be the model for Ryan O'Neal, and "a Vietnam veteran. One of the lucky ones." (I'll say: he spent five months there, well away from the front, as an army PR officer.) So his aides immediately found themselves on the defensive. "The fact is," said Gore spokesman Chris Lehane, "that the Vice-President growing up spent virtually every summer on his family farm."

I think I speak for all my North Country neighbours at the start of another gruelling mud season (the annual snow melt) when I say that the test of a country boy is not whether you spend "virtually" every summer on the farm but whether you can stick every winter. But the Vice-President is unrepentant. "I *did* have a treehouse," he insisted to the children of Sioux Falls Elementary School this week. "I had a really neat treehouse."

Al Gore isn't the first Beltway insider to take a double-bladed axe to his necktie and tasselled shoes and clamber into the old plaid and dungarees. But his rustic routine is peculiarly unconvincing. So, while Bill Clinton's busy bombing the Serbs, Al Gore is busy bombing. Every time he opens his mouth, he blows another huge crater in his poll numbers. Things are so bad he's reduced to getting the President to defend his integrity. "The Vice-President is by nature a reticent person when it comes to talking about his life and his background," said the President. "You want to do it without seeming to toot your own horn too much."

Sound advice. Unlike Mr Gore, Mr Clinton always has someone to hand to toot his horn for him.

Great BBC departures

January 9th 1999
The Spectator

ALMOST EXACTLY six years ago, I was sitting in the BBC studios at Rockefeller Center, just before the start of a weekly talk show I used to host for Radio Four. With the minutes ticking away and the guests arriving, the head of BBC New York was niggling with me over some line in the script he didn't care for - not an unusual occurrence but one which seemed to be distressing to our token American researcher, Nancy. Eventually, she could control herself no longer: "Hey!" she said. "First rule of broadcasting: don't fuck with the talent!"

The otherwise all-British production team looked flummoxed. "The talent" is the American term for what the Corporation calls "the presenter", and at the BBC fucking with them is a way of life. The latest victim is "Feedback" presenter Chris Dunkley, who was informed just before Christmas that, although "Feedback" would be returning in the new year, he wouldn't. I salute Dunkley for declining to finish the remaining two weeks of his stint and quitting on the spot - not like my wimpsville ex-colleagues on "Kaleidoscope", who, after the programme's demise had been announced by the network a year in advance, sat it out on death row in hope of a last-minute reprieve, notwithstanding the fact that the network bigshots had ostentatiously made it known that no presenter of the old clapped-out "Kaleidoscope" would be considered for a slot on the exciting new-look arts show.

Funnily enough, the exciting new-look arts show seems to have plenty of room for the old-look, clapped-out "Kaleidoscope" producers. At the BBC, the problem is always the presenter. So, when Gerry Anderson, of Radio Four's "Anderson Country", was axed, the editor and production team simply renamed the show "The Afternoon Shift", brought in a new presenter - Gerry Afternoon, if memory serves - and carried on as before. Whatever one feels about the tedious Ulster whimsy-monger, surely the real mistake was made by the executive pen-pushers who came up with the bright idea of giving Radio Four's first-ever one-man daily show to a fellow with virtually zero experience on the network.

Still, ask not for whom the Greenwich Time Signal pips, it pips for thee. Four years ago, I too was served with a one-way ticket to Anderson country. A few weeks before the new series of the above-mentioned Saturday-evening New York show was due to start, I was sacked. The only explanation came from Anne Winder, head of magazine programmes, who told my assistant in London that she'd seen a book review I'd done for *The Mail On Sunday* in which, en passant, I'd made a few jokes about BBC New York's recent move to less lavish accommodation in a disused abortion clinic - from which Miss Winder had concluded that I was unhappy with the new office facilities. So the network decided it would be kindest to abort me in the first trimester.

I would have done a Dunkley and refused to host my remaining shows, but, to my surprise, I found I didn't have any. At the time, I also presented the movie edition of "Kaleidoscope" on Wednesdays: for some years, before the start of each quarter, they'd send my assistant a letter enquiring as to when I'd be in London and schedule me accordingly. But, with a horrified gasp, we realised that this time round the quarterly letter hadn't come: my eight years on the show had come to an end without so much as a "Dear Sir or Madam, We regret to inform you..." My cancer of the career, as they call it in American showbiz, had spread to every string on my bow.

It was not an amicable termination. My attorney in New York was obliged to threaten legal action to prevent them using my old signature tune - which probably doesn't seem like a big deal to you, but it gets to the heart of the issue: to what extent is the presenter responsible for the programme's identity? It would seem self-evident that, after 13 years, insofar as "Feedback" has an identity, it derives largely from Dunkley - from his stewardship, reputation, authority, interviewing style. But that's not how Radio Four think of it: they see it, as that old programme title "The Afternoon Shift" suggests, as just a shift - that any old casual labour can be brought in off the street to do.

One recalls, among many examples of classic BBC departures, the late "Start The Week" controversialist Kenneth Robinson. Personally, I always felt that his "controversial reputation" was intended to absolve the rest of the programme from the requirement to be in the least bit interesting. But nothing became his "Start The Week" like the stopping of it. One morning, Robinson turned up to be told his services would no longer be required, and, to add insult to injury, then had to sit through Richard Baker's bland perfunctory thank-you at the end of the week's show. From off-mike, Robinson raged, "It's a bloody disgrace after 17 years."

"Yes, well, there we are," purred Baker, and on came the ten o'clock pips. In the vault at Broadcasting House a few years back, I happened across the recording of the incident: it's a telling comment on the show that the only moment BBC Archives thought worth preserving was the clumsy ejection of a regular contributor. A few weeks later, one of the departmental executives involved in the decision had a new answering-machine message: "I'm too busy to come to the phone right now. But why not call Kenneth Robinson? He could use the work." A "presenter" is so called because he rarely has any future: the real, enduring "talent" is behind the scenes.

Thus, not long afterward, Richard Baker took what he apparently thought was a temporary break from "Start The Week". Like so many before him, he was to discover that BBC holidays come with one-way tickets.

Anyway, a few months after my demise, I was passing through New York and bumped into my old producer, who took me to Starbucks, bought me a cappuccino and offered me a job writing jokes for my replacement, whose opening monologue it seems left something to be desired. "Get lost," I said. "Then people'll say, 'Hey, this new fellow's much funnier than that Mike Stain guy was.' Why would I do that?"

"Take it or leave it," he shrugged. "You're dead at this network. I can't give you away."

But that's the odd thing at the BBC. A presenter is never really dead: no sooner do you slide down some slimy duplicitous executive snake than you're back on the bottom rung of the Corporation ladder heading up. I once mentioned in some column that the colossus of easy listening, Radio Two's urbane charmer David Jacobs, had learned of his own downfall by hearing it announced on the news bulletin immediately preceding his show. As his celebrated Ray Conniff signature tune faded away, the usually unflappable Jacobs sounded very flapped: "Well, I must say, ladies and gentlemen, in 40 years as a professional broadcaster, I've never heard anything quite so..." Etc.

A week after I'd made passing reference to this incident, a letter arrived from Radio Two's deputy supremo indignantly denying that he and his colleagues were a bunch of heartless bastards and explaining that, in fact, things had been arranged so that Jacobs's extinction would be sensitively announced after his show, during the 2 pm news bulletin, and that it had only been announced during the 1 pm bulletin due to an unfortunate clerical error that could happen to anyone.

So David went away for a while. But a year or two later he started creeping back into the schedule, not with his prime-time daily show but with something called "Easy Does It", which was broadcast at 11.43 pm every third Wednesday if the BBC Light Orchestra concert didn't overrun. Life has few aural pleasures more delightful than hearing David's honeyed tones introduce the vocal stylings of Mister Johnny Mathis, but, after what befell him, I can't see why on earth he'd want to get mixed up with these guys again.

Yet the same thing happened to me: I'd barely been lowered into my grave before they began trying to chisel my coffin lid off. At first it was just small things - would I come in and review a Sondheim revival at the Donmar Warehouse? - but perhaps if I stuck it out it might lead to something.

"He's not interested," said my assistant. "After what happened, he'll have no further truck with the BBC. Take him out of your Rolodex."

But they never do. They won't take no for an answer, not now not never. One producer left three increasingly hysterical messages, culminating with "Come on! You know you want to do this!" After a while, we decided there was no point wasting money returning their calls. A week later, a World Service producer left a message at my home in London: "Look, I can tell from the short beeps someone's picking up messages at this number. I don't know who you think you are, not calling us back." Er, someone who's not interested?

In 1997, a young producer from Mentorn, an independent production company, called to say they were "considering" me for a new Radio Four series.

"You're wasting your time," said my assistant. "He doesn't do anything for the BBC."

"Hmm," he said. "Tell me, does he have *any* TV or radio experience?"

"No, no, you've misunderstood," she said. "He *won't* do anything for the BBC."

"Yes, well, I can't promise anything," said the Mentorn man, "but we'll bear him in mind."

A year ago, I was briefly back in Broadcasting House to plug my book. A couple of days after my return to New Hampshire, a letter from James Boyle, the new Radio Four Controller, uncoiled over my fax machine. "I enjoyed hearing you on 'Loose Ends' on Saturday," he wrote. "You sounded as if you enjoyed it too!" He was faxing to offer me the new Radio Four film programme on Saturday evenings and wanted to, as he put it, "give you lunch" at my earliest convenience and introduce me to "the new Radio Four team".

How time flies! A mere three years after being sacked from my Saturday evening show and the network's film programme, I was being offered a film programme on Saturday evenings. I did think of writing back to suggest that perhaps the new-look Radio Four deserved more dramatic innovations. But somehow I never got round to it.

So on it goes: the other day it was a call from some Current Affairs dude demanding that I audition my views on Bill Clinton so he could decide whether I was worth proposing for a spot on some impeachment debate. No one is indispensable - especially when Radio Four apparently has a factory in Glasgow cranking out endless supplies of Scottish women with dreary voices. But why do people do it? If you look closely at the daily schedule, you'll see that what the network actually is is a succession of freelances who've acquired an expertise at some other organisation's expense and are then prevailed upon to give, it away to the BBC for bargain-basement rates.

In the distant days when the Home, Light and Third Programmes enjoyed a broadcasting monopoly, there might conceivably have been some justification for this disadvantageous arrangement. But it's incredible that it survives today. One reason is that, as my old chum and indestructible survivor David Frost always says, Britain doesn't have an independent production sector, only a dependent production sector. "Feedback" is made by Test Bed Productions, whom you'd have thought might have mounted a spirited defence of their man and his excellent track record. But it seems their fear of jeopardising future Radio Four commissions outweighs any loyalty to Dunkley.

Ever since I was a teenager, I've worked on and off for commercial broadcasters in North America, where you live and die according to ratings and revenue - which are at least objective measurements. But so-called public service broadcasting proceeds on nothing more rational than executive whim. At first, my one-man boycott was intended to run only so long as Michael Green remained Controller of Radio Four. But, following the traditional BBC career trajectory, he was soon consigned to the oblivion of early retirement, and I found that my boycott had mellowed into a general contentment at a BBC-free life. If every freelance did as I did, the network would collapse - or, at any rate, degenerate completely into an affirmative-action system for approved regional accents. So I say to Dunkley: take it from me, man; don't be a David Jacobs, eschew "Easy Does It", learn your lesson. It's time the talent started fucking 'em back.

FOOT

Stool

March 1999
The American Spectator

ONE SCENE from the Senate impeachment trial sums up the entire chamber, even though it's not, in any sense, political. It was a Saturday afternoon, a busy one for the Senate pages. Under the cumbersome impeachment procedures, senators had to submit questions to their respective party leaders, who then passed them to the Chief Justice, who then read them out. So the poor old pages were run off their feet ferrying lethal interjections from chief Democrat saboteurs Tom Harkin and Pat Leahy up to their leader Tom Daschle.

The page had barely dropped off Senator Harkin's question when Edward Kennedy, the coughing, heaving, bloated Senator from Massachusetts, called him over.

"Ah-ha!" I thought, from up in the gallery. "Ted's going to spring a surprise question on the prosecutors." The page silently padded over to the Senator's seat in the back, Ted whispered to him, and the page made his way to the end of the row, then worked his way along the row in front, squeezing past senators until he was directly facing Ted's desk. He then dropped to his knees. "My God!" I fretted. "Don't tell me he's going to give him the full Monica." But no; instead, he leaned under the Senator's desk and adjusted Ted's footrest by an inch and a half.

The guy was squeezing his way back past the other senators, Pat Leahy having some other brazenly irrelevant question he needed taking to the minority leader, when Ted signaled him back. The page turned around, squeezed past Senator Graham of Florida yet again and dropped to his knees to move Ted's footrest another smidgeonette. He then rushed off to pick up Senator Leahy's note. Senator Kennedy didn't thank him.

It would be ludicrous to expect Teddy Kennedy to be concerned about the civil rights of someone like Paula Jones: The President had a "problem," he got his people to "fix" it - just like Kennedys have been doing in Massachusetts for three generations. Nor would he trouble himself much over Bill and Monica's "consensual relationship": It was very consensual. Not only did she eagerly service Clinton, but she consented to let the President place her in legal jeopardy and conscript her to his conspiracy - just as obliging nobodies have been willing to do for the Kennedys for years. And though few members of the chamber are as grand as the senior Senator from Massachusetts, his values have seeped into its bloodstream: Senators seem all but incapable of lowering their sights to the humble citizens - the Paula Joneses and Kathleen Willeys - clustered round their footstools.

One of the most admirable qualities of this President is his very personal commitment to diversity. He digs chicks - all sizes, all shapes, all ages: airhead Valley Girls like Monica, twangy trailer trash like Paula, elegant widows like Kathleen, beauty queens, black hookers, you name 'em, he's hit on 'em. But, curiously enough, young or old, rich or poor, smart or dumb, none of them seems to fall into any category of womanhood that Democratic senators give a hoot about.

TOE

Job

August 20th 1992
The Evening Standard

FOLLOWING the riveting display by the Duchess of York, many readers have written enquiring about correct etiquette during what *The Sun* calls "toe jobs".

For example, is it acceptable to suck toe on the first date?

Yes, but the shyer sort of girl may prefer to keep her stockings on and men should always respect a woman's right to shoes. If your partner wears Crimplene socks, however, you may find the static electricity adds to the sensation.

Size isn't everything, either. A hangnail can be just as much fun as a well-hung nail, so don't confine yourself to the big toe: Remember, as the old song says, "I Saw Mommy Kissing Santa Claus Underneath The Little Toe Last Night".

Several people have complained that sex manuals devote very little space to toe-sucking (although John Gummer is issuing a leaflet on foot and mouth disease). So it's worth noting the basic steps:

1) wear loose-fitting clothing;

2) be considerate - always swallow a couple of Odour-Eaters beforehand;

3) grasp foot firmly with both hands, raise to mouth and then, as the old song says, "Tip Toe Through The Two Lips". If you're in a car and have a sudden desire to get your toe away, remember to park in a toe-away zone. In all other areas, wear a Denver Boot.

TOE JOB

If you are with a member of the Royal Family or a senior Conservative politician, he or she may wish to keep the foot discreetly veiled to avoid being recognised (this is known as "travelling incognitoe"). Even with sophisticated types, do not attempt to inject intravenous drugs through your feet as you could end up "coma-toes".

Lastly, always bear in mind that personal hygiene is extremely important, so do tidy up afterwards: even the most zealous toe-sucker doesn't want to sleep in a bed of nails (see my enclosed clippings).

HEART
and soul

September 10th 2001
The National Post

D R CHRISTIAAN Barnard was nothing if not vain and it must have bothered him, wherever he is now, that his passing drew so little comment. ABC News rated his death below that of the ten-year old Virginia shark victim and barely any more noteworthy than that of Troy Donahue, hormonal heartthrob of *A Summer Place*, the fevered blockbuster of four decades ago. Fleet Street ran a few half-hearted pieces pointing out what a racist bully the doctor was, yet even they saw his reaction to the end of apartheid (he decided to leave South Africa) as a paradox, an example of how even the most compassionate benefactors to mankind have their dark side.

As with *A Summer Place*, it's hard from the vantage of 2001 to realize how large Dr Barnard once loomed. His legacy can be found not just in the thousands to whom he gave new life but in the millions more on whom he had the most profound cultural influence. Before the good doctor, the heart was not just an organ, not just a body part, but an intimate essence of our identity:

Heart And Soul
I fell in love with you...

When I give my heart
It will be forever...

Most of us know of at least one example among friends of friends, or friends of friends of friends, where after 60 years of marriage one partner expires and the spouse follows a week later and is said to have died of "a broken heart". Not a broken heart in the Dick Cheney sense, a matter of stents and bypasses, but something more mysterious and fundamental to our sense of ourselves. No-one ever figured the world needed a song called "Kidney And Liver/I fell in love with you..."

Barnard's success was due not only to his first transplant (1967) but also in the way he negotiated a widespread unease about the procedure. The patient's new heart is, after all, someone else's old heart. Dr Juro Wada was not so canny. A year after Barnard, he performed Japan's first heart transplant, but, lacking the South African's PR skills and more fortuitous political climate, Dr Wada found himself investigated for murder over the precise point at which the organ had been removed from its original owner. He was never indicted, but the furore so discredited the entire procedure that it was not until 1999, 31 years later, that anyone got around to performing Japan's second heart transplant.

Barnard's favourite dinner-party anecdote concerned one of his American customers, a wealthy southerner who'd flown to Groote Schuur to await the next car accident and a potential donor. Looking up from his bed, he sternly warned Barnard's colleague, "Sir, I don't want no black heart." The doctor re-told this anecdote for 30 years, the southern drawl growing ever more exaggerated. Barnard was right to find his patient's position absurd: either the heart is just a pulsing mass of ventricles and arteries or it's something more particular. But if you object to a black heart why not to the white heart of a serial killer or a bow-tied twit on the "Antiques Roadshow" or a teenybopper who's nuts for some airbrushed ninny? There are science absolutists and moral absolutists – those who think all abortions are wrong, all embryo stem-cell research is objectionable – but Barnard understood that in between the vast majority is like that southern client, its views on such issues a contradictory mess of prejudice and self-interest and sentimentality which a savvy operator can easily drive a truck through.

Apparently, the next big transplant thing will be heads. The initial beneficiaries would be those paralysed from the neck down, who, despite perfectly healthy heads, often die prematurely because of multiple-organ failure. If one were to transplant their heads on to healthy bodies, they'd still be paralysed, but their life expectancy would be dramatically improved. Christopher Reeve has been in the forefront of demands for stem-cell research because he has pledged – to the irritation of others with his condition – to walk again and is anxious for science to catch up with his dreams. But it's not difficult to imagine him repositioning his ambitions and championing increased funding for head transplants: it is, after all, different only in degree from what many of his fellow thespians – Pamela Anderson, Anna Nicole Smith – have already undergone.

And, once we get used to the idea of head transplants for the paralysed, why not for the rest of us? I've often felt, for example, that my extreme right-wing views would be far more saleable if I had Naomi Klein's body. But how would a Naomi Steyn work in practice? At the Summit of the Americas perimeter fence, would my heart be telling me to join the ski-masked anarchists and start lobbing concrete while my head was tugging me over to the other side to grab the water cannon and start hosing down Maude Barlow?

What, in other words, comprises the essence of me? The Barnard view of man is like the joke about the old rustic and the axe he's had for 70 years. He's replaced the blade five times and the handle six times, but to him it's still the same old axe. In yesterday's *Toronto Sun*, Peter Worthington quoted his own surgeon, Tirone David: "Basically, the heart is just a pump, and I'm just a plumber." In that case, is the head the house and the body the foundation? Or is the body the house and the head just some semi-finished attic space where you store a lot of old junk you don't really need?

Heart and soul: Man has achieved mastery over the former, so it was inevitable that he would attempt to finish the job. Stem-cell harvesting, cloning, egg sales on the Internet: what are these things *for*? In that respect, Dr Barnard's life is as instructive as his work: he used his celebrity to live it up, screw broads, get free Italian hand-made suits. He was a hedonist and narcissist, and why wouldn't he be? For if man is now his own god, then one should serve oneself as devoutly as our superstitious fathers served their unseen Lord. So Barnard graduated from heart transplants to lending his name and prestige to a quack anti-aging cream. Which is what it's about, isn't it? Man's victory over death – not in the sense in which our ancestors believed, the certainty of eternal life in the unseen world, but in the here and now. Who knows what the great beyond's like? What are the chicks like? What are the suits like? Better to conquer death here on earth.

We know what stem-cell research isn't for. It isn't so that, say, a mother carrying a foetus in whom birth defects have been detected can be told by her doctor, "Well, with all this new stem-cell research, he'll be able to live a reasonably normal life." No, that baby will still be aborted. But in thrall to Narcissus we'll be able to live forever and clone even more perfect versions of ourselves, because we know now that we are not hearts and souls but cells and genes and DNA – a redefinition of ourselves begun 34 years ago by Christiaan Barnard.

Will it change our essential humanity? Who cares? "This warfare against nature must end once and for all," declared Michael Fox – not the great Canadian actor and stem-cell activist but the Humane Society's Senior Scholar for Bioethics. "We are very clever little simians, aren't we? Manipulating the bases of life and thinking we're little gods." But he wasn't talking about embryonic stem cells or cloning. He was speaking at a rally against genetically-modified foods. If you clone genetically-modified pigs so that their body parts can be harvested for humans – as Edinburgh's PPL Therapeutics successfully did last year – Mr Fox is entirely relaxed. But if you grow a genetically-modified zucchini, the stormtroopers of the enviro-left will burn your crops and blow up your factory. We don't know the long-term effects of a genetically-modified cucumber, fret the Foxes of the world. But on the long-term effects of genetically-modified humans they're positively insouciant.

Poor old Mary Shelley got it so wrong. Picture Castle Frankenstein today, the burgomeister and the villagers with the pitchforks and torches marching up the hill as the seven-foot guy with the bolt through the neck staggers back from the kitchen garden with a little light salad for the Baron. "Oh, my God!" they cry. "Look at the size of that tomato!"

APPENDIX

...and other things red wine is good for

APPENDIX

April 21st 2001
The Daily Telegraph

Our Medical Correspondent, Dr Mac Onrouge of Oddbins, answers your queries:

Doctor, I was fascinated by the story in The Daily Telegraph *this week that red wine, aside from preventing cancer, heart disease, Alzheimer's, skin deterioration and sleeping sickness, is now thought to be a potential cure for Aids. Can this really be true?*

Absolutely. Red wine is good for anything. It's the leech of the 21st century. But it's still very important to practise "safe sex". For example, if you meet a fetching young man called Kevin in a bar on Old Compton Street, but you're not sure of his sexual history, I'd recommend the 1998 Pinot Noir Val St Grégoire. Mmm. Light and fruity with a perfumed nose. But enough about Kevin, what about the wine? Well, slug it back. But, if you're thinking of going all the way, why not order a second bottle? Maybe a 1995 Coonawarra. Mmm. Smooth, dark, mulberry-hued and huge down under...

Yes, yes, enough about Kevin. Should we use a condom?

Certainly. There's nothing to be embarrassed about. Just remove the condom from the packet and, using your thumb and forefinger, roll it down the neck of the open bottle. That keeps the wine nicely sealed so it'll still taste great after you've finished having sex!

Doctor, I'm due to have my appendix removed on Monday. Is this a simple procedure?

Couldn't be easier, as I'll demonstrate with the assistance of my nurse, Francine. Scalpel.
    Scalpel.
Clamp.
    Clamp.
Corkscrew.
    Corkscrew.
A 1997 Mâcon Superieur.
    Sorry, Doctor, we only have the '98.

344

Whatever, Francine. Just pour it in the incision. Mmm. Soft and supple with just a hint of sutures.

Doctor, I've got two on the aisle for My Fair Lady *this week, but I'm very concerned about Martine McCutcheon's health. If she comes over all queer during "Wouldn't It Be Luverly?", I'm thinking I ought to have a St Émilion with me, just in case I need to pour it over her.*

No, no, no. Never, never do that.

You mean I should leave it to a qualified medic?

No, you should leave it to a qualified Médoc. If anything happens at the theatre, just stand up calmly and say: "Is there a wine merchant in the house?" In crisis situations, we know exactly what life-saving treatment to administer for each star. Go on, test me.

Liza Minnelli?

Life is a Cabernet, old chum.

Jason Donovan?

Joseph and the Amazing Technicolor Dreamcôtes du Rhône.

Er, Lassie?

Easy. A 1991 Chianti Collie Fiorentini and, if she's very good, a Côtes de Beaune.

That's all very well, Doctor. But suppose you're off-piste at Val d'Isère and you fall awkwardly and tear a ligament, with no one else around. What then?

That's why, whenever I ski, I always strap a case of the 1994 Cave de Tain l'Hermitage Nobles Rives Crozes-Hermitage to my back. After the fall, rub the wine into the ligament, tie the empty bottle behind the knee to support the injured leg, and gently proceed down the slope until you come to the nearest off-licence.

Doctor, Tony Blair promised he'd reduce NHS waiting lists. But I'm

87 and I've been waiting for a routine Bulgarian merlot for almost two years now.

Well, it may be time for you to think about going private.

That's fine for Tony and Cherie, popping in for a same-day Chateau Petrus. But I can't afford that. I need a hip replacement and they said I'd be looking at an '82 Mouton-Rothschild for £3,800.

Utter nonsense. What we have now are these new non-brand name generic drugs - or, as we call them, *vins du table*. They're a little rough, but they're perfectly safe for routine procedures. Come in tomorrow, we'll take out the hip and replace it with a large carafe of house red for £2.99.

Doctor, every morning I take a glass of 1995 Montes de Ciria Rioja for my HIV, a glass of 1997 Chateau de Laurens Faugeres for my Alzheimer's, a glass of 1992 Chateau de Rives Bordeaux Superieur for my lumbago, a glass of 1989 Grao Vasco Garrafeira for my carpal tunnel syndrome, and half a bottle of Valpolicella to control my back hair. Am I at risk from any side effects?

Aside from a slight tendency to drive across the central reservation on the way to work, there are no known side effects.

Doctor, is there now any illness left that red wine can't cure?

I used to think so. Then one day I was performing complex spinal surgery on a patient and the spliced bone ends just weren't gelling. Frankly, I was out of ideas. And then suddenly my supplier burst through the door yelling, "*Le* bone gel aid *nouveau est arrivé.*" An unpretentious little wine, but that man is alive today because of it.

My God, that's a moving story.

I know. I used it in my lecture at the Eye, Ear, Nose and Throat Hospital. There wasn't a dry throat in the house.

ALSO BY MARK STEYN

The Face
Of The Tiger

AND OTHER TALES FROM
THE NEW WAR

"The only change that occurred on September 11th was a simple one. When Osama bin Laden blew up the World Trade Center, he also blew up the polite fictions of the pre-war world..."

In this collection, Mark Steyn considers the facts and fictions of the post-9/11 world – war and peace, quagmires and root causes, from "the brutal Afghan winter" to the "searing heat of Guantanamo", from the axis of evil to the "axels of evil", the heroes of Flight 93 to the death of Osama bin Laden.

This volume rounds up Steyn's columns from across the world, including his take on pop, movies and plays since September 11th – from Ridley Scott's *Black Hawk Down* to Neil Young's "Let's Roll" to Ovid's *Metamorphoses* – and some satirical flights of fancy: enjoy an afternoon hanging upside down in the bondage dungeon with Robert Fisk or sautéing scorpions at the back of the cave with Mullah Omar.

CANADA	UNITED STATES	UNITED KINGDOM
$29.95	$19.95	£12.50

ACKNOWLEDGMENTS

"The Air That I Breathe" (1974) by Albert Hammond and Michael Hazlewood © EMI
April Music

"Before The Music Ends" (1979) by Gordon Jenkins © Warner Chappell Music

"Black Coffee" (1948) by Sonny Burke and Paul Francis Webster © Sondot Music/
Webster Music

"Crazy Rhythm" (1928) by Roger Wolfe Kahn, Joseph Meyer and Irving Caesar © Warner
Chappell Music/Irving Caesar Music/JoRo Music

"Fly Me To The Moon" (1954) by Bart Howard © Hampshire House Publishing

"Fools Rush In" (1940) by Johnny Mercer and Rube Bloom © Bregman, Vocco & Conn/
Mercer Music c/o Warner Chappell Music

"Hamlet" (1949) by Frank Loesser © BMG Music

"Have You Seen My Childhood?" (1995) by Michael Jackson © Mijac Music

"Heart And Soul" (1938) by Hoagy Carmichael and Frank Loesser © Famous Music

"I Get A Kick Out Of You" (1934) by Cole Porter © Warner Chappell Music

"If I Only Had A Brain" (1939) by Harold Arlen and E Y Harburg © EMI Feist Catalog

"The Joker" (1973) by Steve Miller, Ahmet Ertegun and Sonny Curtis © Sailor Music/Hill
& Range Songs c/o Warner Chappell Music

"Mama Will Bark" (1951) by Dick Manning

"My Way" (1968) by Claude François, Jacques Revaux, Gilles Thibault and Paul Anka
© Chrysalis Standards

"Rudolph The Red-Nosed Reindeer" (1949) by Johnny Marks © St Nicholas Music

"Santa Claus Is Coming To Town" (1934) by J Fred Coots and Haven Gillespie © EMI
Feist Catalog/Haven Gillespie Music

"Smoke Rings" (1932) by Gene Gifford and Ned Washington © Music Sales/EMI
Music/Old Acct

"These Foolish Things" (1935) by Holt Marvell, Jack Strachey and Harry Link © Bourne
Music

"When I Fall In Love" (1952) by Victor Young and Edward Heyman © Intersong
USA/Victor Young Publishing

"You're Getting To Be A Habit With Me" (1933) by Harry Warren and Al Dubin
© M Witmark and Sons

Cover and contents pages: *Vitruvian man* by Leonardo da Vinci

MARK STEYN RETURNS...

...in this Sunday's Chicago Sun-Times *and* Sunday Telegraph;

...in Monday's Irish Times, New York Sun *and* Washington Times;

...in Tuesday's Daily Telegraph;

...in Wednesday's Jerusalem Post;

...in Friday's Spectator;

...in the next National Review, Atlantic Monthly *and* New Criterion;

...and at SteynOnline.com